STEALING FIRE

Also by Steven Kotler

Abundance:
The Future Is Better Than You Think

Bold:
How to Go Big, Create Wealth, and Impact the World

Tomorrowland:
Our Journey from Science Fiction to Science Fact

The Rise of Superman:
Decoding the Science of Ultimate Human Performance

West of Jesus:
Surfing, Science, and the Origins of Belief

A Small Furry Prayer:
Dog Rescue and the Meaning of Life

The Angle Quickest for Flight

STEALING FIRE

How Silicon Valley, the Navy SEALs, and Maverick Scientists Are Revolutionizing the Way We Live and Work

STEVEN KOTLER & JAMIE WHEAL

An Imprint of WILLIAM MORROW

The information contained within this book is strictly for educational purposes. This book is not intended to be a substitute for the medical, legal or pastoral advice of licensed professionals. If you wish to apply ideas contained herein, you are taking full responsibility for your actions. *Caveat Lector!*

 For information, address HarperCollins Publishers, 195 Broadway, New York, NY 10007.

HarperCollins books may be purchased for educational, business, or sales promotional use. For information, please email the Special Markets Department at SPsales@harpercollins.com.

FIRST EDITION

Designed by Michelle Crowe

Library of Congress Cataloging-in-Publication Data has been applied for.

ISBN 978-0-06-242965-0

17 18 19 20 21 LSC 10 9 8 7 6 5 4 3 2 1

To Julie, Lucas, and Emma, sine qua non.
J.W.

To William James, you got there first.
S.K.

CONTENTS

PART TWO: THE FOUR FORCES OF ECSTASIS

PART THREE: THE ROAD TO ELEUSIS

INTRODUCTION

The Never-Ending Story

Some revolutions begin with a gunshot, others with a party. This one kicked off on a Friday night in downtown Athens, in 415 BCE. Alcibiades, a prominent Greek general and politician, had invited a small circle of friends to his villa for what was to become one of the more infamous bacchanals in history. Hooded in the stolen robes of a high priest, Alcibiades swept down his marble staircase, recited a forbidden incantation, and produced an ornate decanter. Carefully, he poured a single shot of a dark liquid into each guest's glass. A few more words, an exuberant cheer, and everyone drained their cups.

In less than an hour, the effects took hold. "Fears, terrors, quiverings, mortal sweats, and a lethargic stupor come and overwhelm us," the historian Plutarch later recounted. "But, as soon as we are out of it, we pass into delightful meadows, where the purest air is breathed, where sacred concerts and discourses are heard; where, in short, one is impressed with celestial visions."

By sunup, those visions had faded, replaced by repercussions in the real world. Alcibiades's illicit party kicked off a chain of events that would prompt him to flee Athens, dodge a death sentence, betray his government, and set in motion the trial and execution of his beloved teacher, Socrates.

Famously handsome, eloquent, and ambitious, Alcibiades's faults were as plentiful as his gifts. He offered sex to Socrates in exchange for the philosopher's deepest secrets. Before his wife could divorce him for womanizing, he dragged her out of court by her hair. Politically, he played both sides against the middle, and his only true allegiance was to his career. So when his rivals got wind of that scandalous evening, they ratted him out to the highest Athenian court for stealing "*kykeon,*" the sacred elixir he'd shared with his guests. He was tried in absentia for a crime punishable by death—blaspheming the Mysteries.

And not just any mysteries; the Eleusinian Mysteries, a two-thousand-year-old initiatory ritual that had an outsize impact on Western philosophy and counted some of Greece's most famous citizens among its elect. Foundational notions like Plato's world of forms and Pythagoras's music of the spheres were informed by these rites. "Our Mysteries had a very real meaning," Plato explained, "he that has been purified and initiated [at Eleusis] shall dwell with the gods." Cicero went further, calling the rites the pinnacle of Greek achievement: "Among the many excellent and indeed divine institutions which . . . Athens has brought forth and contributed to human life, none, in my opinion, is better than the Mysteries. . . . In [them] we perceive the real principles of life, and learn not only how to live in joy, but also to die with better hope."

In more contemporary terms, the Eleusinian Mysteries were an elaborate nine-day ritual designed to strip away standard frames of reference, profoundly alter consciousness, and unlock a heightened level of insight. Specifically, the mysteries combined a number of state-changing techniques—fasting, singing, dancing, drumming, costumes, dramatic storytelling, physical exhaustion, and *kykeon* (the substance Alcibiades stole for his party)—to induce a cathartic experience of death, rebirth, and "divine inspiration."

And so powerful was this experience and so significant were those insights that the Mysteries persisted for more than two thousand years. A lesser ritual would have fizzled or, at least, become an empty gesture devoid of its original power. Eleusis, historians tell us, endured time and turmoil for a couple of key reasons: First, initiates kept the mystery in the Mystery—disclosing any of its secrets, as Alcibiades did, was a capital offense. And second, *kykeon,* that dark liquid at the heart of the ritual, packed one hell of a punch.

For anthropologists, uncovering the ingredients of *kykeon* has become a Holy Grail kind of quest. It ranks right up there with decoding *soma,* the ancient Indian sacrament that inspired Aldous Huxley's groupthink happy drug in *Brave New World.* Swiss chemist Albert Hofmann and Harvard-trained classicist Carl Ruck argued that the barley in *kykeon* might have been tainted with an ergot fungus. This same fungus generates lysergic acid (LSA), a precursor to the LSD that Hofmann famously synthesized in his Sandoz pharmaceutical lab. When consumed accidentally, ergot prompts delirium, prickly limbs, and the hallucinations known as "St. Anthony's fire." When taken on purpose, within the context of an intensive initiatory ritual, you have all the ingredients of a highly effective ecstatic technology—so effective (and, presumably, so enjoyable) that Alcibiades was willing to risk his life to steal it for a party.

All of which is to say, as far back as we can trace Western civilization, buried among the stories that bore schoolchildren to tears, we find tales of rebel upstarts willing to bet it all for an altered state of consciousness. And this isn't an isolated incident. It's just an early indicator of a perennial pattern, hidden inside of history, tucked among the names and dates we know so well.

At the center of this dynamic sits the myth of Prometheus, the original upstart rebel, who stole fire from the gods and shared it with humankind. And he didn't just steal a book of matches, but also the power to seed civilization: language, art, medicine, and technology. Enraged that mortals would now have the same power as the gods, Zeus chained Prometheus to a rock, letting eagles rip out his innards for eternity.

This story has continued to repeat itself throughout the ages. Typically, a rebel, seeker, or trickster steals fire from the gods. It can take the form of a potent celebratory rite, a heretical new scripture, an obscure spiritual practice, or a secret, state-changing technology. Whatever the case, the rebel sneaks the flame out of the temple and shares it with the world. It works. Things get exciting. Insights pile up. Then, inevitably, the party gets out of hand. The keepers of law and order—call them the priests—spot the hedonistic blaze, track down the thief, and shut down the show. And so it goes, until the next cycle begins.

Stealing Fire is the story of the latest round in this cycle and, potentially, the first time in history we have a chance for a different ending. It's the story of an entirely new breed of Promethean upstart—Silicon Valley executives, members of the U.S. special forces, maverick scientists, to name only a few—who are using ecstatic techniques to alter consciousness and accelerate performance. And the strangest part? It's a revolution that's been hiding in plain sight.

Accidental Prometheans

If a revolution is the kind of thing you can stumble upon, then we—your authors, Steven and Jamie—stumbled upon this one a few years ago. And really, we should have seen it coming.

That's because at the Flow Genome Project we study the relationship between altered states and peak performance, focused primarily on the experience known as *flow*. Defined as an "optimal state of consciousness where we feel our best and perform our best," flow refers to those "in the zone" moments where focus gets so intense that everything else disappears. Action and awareness start to merge. Our sense of self vanishes. Our sense of time as well. And all aspects of performance, both mental and physical, go through the roof.

Scientists have known about the relationship between flow and peak performance for more than a century, but a real understanding of this relationship has been slow in coming. The main problem was conflicting motivations. The people really good at finding flow, mostly artists and athletes, were rarely interested in studying it. And the people interested in studying flow, primarily academics, were rarely good at finding it.

"We founded the Flow Genome Project in an attempt to solve this problem. Our goal was to take a multidisciplinary approach to mapping the neurobiology of flow, and then open-source the results. But to do this, we had to establish a common language around these states. So Steven wrote *The Rise of Superman,* a book about the neuroscience of peak performance and action sports.

Following the book's release, we found ourselves talking flow with a wider and wider assortment of people. What began as meetings with individuals and organizations with a vested interest in high-stakes competition—professional athletes and the military—expanded into Fortune 500 companies, financial organizations, tech firms, health-care providers, and universities. The idea that nonordinary states of consciousness could improve performance was spreading out of the extreme and into the mainstream.

But what caught our attention were the conversations we were having *after* those presentations. On too many occasions to count, people would pull us aside to tell us about their clandestine experiments with "ecstatic technologies." We met military officers going on monthlong meditation retreats, Wall Street traders zapping their brains with electrodes, trial lawyers stacking off-prescription pharmaceuticals, famous tech founders visiting transformational festivals, and teams of engineers microdosing with psychedelics. In other words, everywhere we went, someone was trying to steal the *kykeon.*

We wanted to know precisely where this trend was originating and exactly how these leaders were altering their mental states to enhance performance. So we lit out on the trail of these modern-day Prometheans. Over the last four years, this journey has led us all over the world: to the Virginia Beach home of SEAL Team Six, to the Googleplex in Mountain View, to the Burning Man festival in Nevada, to Richard Branson's Caribbean hideaway, to luxurious dachas outside Moscow, to Red Bull's headquarters in Santa Monica, to Nike's innovation team in Portland, to bio-hacking conferences in Pasadena, to private dinners with United Nations advisers in New York. And the stories that we heard stunned us.

In their own ways, with differing languages, techniques, and applications, every one of these groups has been quietly seeking the same thing: the boost in information and inspiration that altered states provide. They are deliberately cultivating these states to solve critical challenges and outperform their competition. It isn't just grit, or better habits, or longer hours that are separating the best from the rest. To hear these trail-blazers tell it, the insights they receive in those states are what make all the difference. And unlike in earlier, more guarded eras, today they're openly talking about their adventures. The ecstatics are coming out of the closet.

Put all these experiences together and it's beginning to seem like a Promethean uprising. Advances in science and technology are giving us unprecedented access to and insight about the upper range of human experience, arguably the most controversial and misunderstood territory in history. Around the world, revelers, soldiers, scientists, artists, entrepreneurs, technologists, and business leaders are leveraging these insights for a common goal: a glimpse above the clouds. First in isolation, then in increasing numbers, and now, if you know where to look, virtually everywhere you look. We are witnessing a groundswell, a growing movement to storm heaven and steal fire. It's a revolution in human possibility.

And this is a book about that revolution.

PART ONE

The Case for Ecstasis

The alternative is unconsciousness, the default setting, the rat race, the constant gnawing sense of having had, and lost, some infinite thing.

—*David Foster Wallace*

CHAPTER ONE

WHAT IS THIS FIRE?

The Switch

One of the hardest parts of being a Navy SEAL isn't knowing when to shoot; it's knowing when *not* to shoot. And we know why. If you put a dozen guys in a dark room and arm them with automatic weapons, somebody's going to blink. Or twitch. Then it's game on. That's what made capturing Al-Wazu such a challenge: more than anything else, the SEALs needed him alive.

It was late September 2004, at a forward operating base in the northeastern corner of Afghanistan. A couple of dozen members of the elite SEAL Team Six, or, in their preferred parlance, DEVGRU, were stationed there, gathering intelligence and staging missions. Some six months prior, a radio operator had noticed a spike in Wazu chatter. Perhaps he was hiding in the woods to the south of them. Possibly he was in the mountains to the north. Then the rumors turned into facts. Wazu actually was in the woods *and* the mountains, holed up in an alpine forest some seventy miles west of their current position.

For the SEALs, this wasn't good news. The terrain to the west was high desert—lonely, barren, and rough. Not enough cover for a stealth mission. Under these conditions, there was no way to get in without a firefight; no guarantee they could capture Wazu alive.

Though he was once a midlevel player, Al-Wazu's notoriety had skyrocketed after he'd pulled off a feat no other Al-Qaeda operative had accomplished: an escape from an American detention center. This single act elevated him to the upper echelons of the organization, earning him a band of committed followers and that ultimate jihadi honor: a personal letter of commendation from Osama bin Laden.

Ever since, Wazu had been busy: recruiting, raiding, and killing. That's why the SEALs needed him alive. His value as an intelligence asset had quadrupled. There was enough in his head to take down most of the remaining cells in the area. Plus, the SEALs wanted to send a message.

And that day in September, they got their chance. The radio call came in the afternoon: Al-Wazu was on the move. He'd come out of the woods and down from the mountains. He was heading straight for them.

For the SEALs, this changed everything. With a moving target, the variables multiplied exponentially. Anything could happen. The team got together and combed through the mission. Contingency plans were put into place, details were committed to memory. Day turned into night and night rolled on.

They only had five hours until dawn, and still no target. The SEALs needed the darkness. Their mission got much more complicated during the day. There were more people awake and more traffic on the roads and too many ways a suspect could disappear into a crowd.

Then, after all that waiting, they suddenly had a target. Al-Wazu had stopped. Only a few hours of darkness remained and the SEALs couldn't believe their luck. He'd holed up less than a mile from their current position—they could literally walk to the op.

Commander Rich Davis (for security, not his real name) wasn't sure it was luck. As the leader of this unit, he knew how badly his men wanted Al-Wazu. They were keyed up. A mile hike wasn't much. Davis would have preferred a three-hour uphill slog. Three hours wouldn't tire them out, but it might calm them down. Might help them focus. Might help them *merge.*

The Greeks had a word for this merger that Davis quite liked—*ecstasis*—the act of "stepping beyond oneself." Davis had his own word as well. He called it "the switch," the moment they stopped being separate men with lives and wives and things that matter. The moment they became, well, there's no easy way to explain it—but something happened out there.

Plato described ecstasis as an altered state where our normal waking consciousness vanishes completely, replaced by an intense euphoria and a powerful connection to a greater intelligence. Contemporary scientists have slightly different terms and descriptions. They call the experience "group flow." "[It's] a peak state," explains psychologist Keith Sawyer in his book *Group Genius,* "a group performing at its top level of ability. . . . In situations of rapid change, it's more important than ever for a group to be able to merge action and awareness, to adjust immediately by improvising."

Whatever the description, for the SEALs, once that switch was flipped, the experience was unmistakable. Their awareness shifted. They stopped acting like individuals, and they started operating as one—a single entity, a hive mind. In the high-stakes hot zone that is their job, this collective awareness is, as Davis says, "the only way to get the job done."

And isn't that peculiar? It means that on the night in question, during a critical mission to capture and not to kill, an altered state was the only thing standing between Al-Wazu and a preemptive double tap to the chest. As isolated individuals, with fingers on the trigger, someone was bound to twitch. But as a team, thinking and moving together? Intelligence got multiplied, fear divided.

The whole wasn't just greater than the sum of its parts; it was smarter and braver too. So Commander Rich Davis wasn't just hoping they'd flip the switch that evening; he was banking on it.

"More than any other skill," he explains, "SEALs rely on this merger of consciousness. Being able to flip that switch—that's the *real* secret to being a SEAL."

The High Cost of Ninja Assassins

It costs $25,000 to turn an average Joe into a combat-ready U.S. Marine. SEALs, meanwhile, cost a lot more. Estimates for eight weeks of Navy basic training, six months of underwater demolition training, six months of advanced skills training, and eighteen months of predeployment platoon training—that is, what it takes to get a SEAL ready for combat—total out to roughly $500,000 per head. Which is to say, the Navy SEALs are among the most expensive collections of warfighters ever assembled.

And that's just the cost of training garden-variety ninja assassins. Making it to the elite DEVGRU unit requires first rotating through several other SEAL teams (there are nine in total). As it costs about $1 million a year to keep a frogman in the field, and these rotations take a couple of years to complete, add roughly another $2.5 million to the tally. Finally, there are additional months of hostage rescue training, which is DEVGRU's specialty, at somewhere north of $250,000 per. All in, those couple dozen men under Rich Davis's command, the SEAL unit charged with capturing, not killing, Al-Wazu, were an exceptionally well-oiled $85 million machine.

So what are U.S. taxpayers getting for their money?

A decent place to start is with the job description itself, or rather, the lack of one. SEALs are multitasking multitools. As their official website explains: "There is no typical 'day at the office' for a Navy SEAL. SEALs constantly learn, improve and refine skills working with their teammates. Their office not only transcends the elements of Sea, Air and Land, but also international boundaries, the extremes of geography and the spectrum of conflict."

The technical term SEALs use to describe these conditions is *VUCA*—Volatile, Uncertain, Complex, and Ambiguous. Prevailing over this type of chaos requires an astounding level of cognitive dexterity. As Rich Davis explains: "the most expensive part of these already expensive warfighters is the three pounds of gray matter resting inside their skulls."

Of course, this isn't how we normally think of SEALs. What we know best about these special operators is how hard they train their bodies, not their minds. Hell Week, for example, the kickoff to their infamous selection process, is five and a half days of nonstop physical exertion and radical sleep deprivation that routinely breaks world-class athletes. But even this crucible is more about brain than body. As SEALFit founder Mark Divine recently told *Outside* magazine, "[T]raining is designed to find the few who have the mental toughness needed to become a SEAL."

"*Grit*" is the term psychologists use to describe that mental toughness—a catch-all for passion, persistency, resiliency, and, to a certain extent, ability to suffer. And while this is accurate—SEALs are gritty as hell—it's only part of the picture. Grit only refers to individual toughness, and the secret to becoming a SEAL has everything to do with team. "At every step of the training," says Davis, "from the first day of BUD/S (Basic Underwater Demolition/SEALs) through their last day in DEVGRU, we are weeding out candidates who cannot shift their consciousness and merge with the team."

On the surface, of course, this seems ridiculous. "*Ecstasis*" is the antecedent for "*ecstasy,*" which, if you can get beyond the club drug references, describes a profoundly unusual state, an experience far beyond our normal sense of self, and definitely not a term traditionally associated with elite special forces. It certainly doesn't show up in the recruiting brochures.

Yet everything we consider SEAL training is actually a brutal filtration system that, beyond the obvious tactical skills and physical perseverance, sorts for exclusively one thing: Does an operator, with his back against the wall, retreat into himself, or merge with his team? This is why they relentlessly emphasize "swim buddies" (the partner you can never leave behind, no matter what) in basic training. Why, even on deployment in Afghanistan—where there's not a body of water for thousands of miles—they still have "swim buddies." It's also how they separate good from great in the fabled Kill House, their specially designed hostage rescue training facility, where they measure a team's ability to move as one by the millimeter, where success requires an almost superhuman collective awareness.

"When SEALs sweep a building," says Rich Davis, "slow is dangerous. We want to move as fast as possible. To do this, there are only two rules. The first is do the exact opposite of what the guy in front of you is doing—so if he looks left, then you look right. The second is trickier: the person who knows what to do next is the leader. We're entirely nonhierarchical in that way. But in a combat environment, when split seconds make all the difference, there's no time for second-guessing. When someone steps up to become the new leader, everyone, immediately, automatically, moves with him. It's the only way we win."

This "dynamic subordination," where leadership is fluid and defined by conditions on the ground, is the foundation of flipping the switch. And, even back when team leaders understood it far less than they do today, establishing this foundation was a top priority. "The Navy's caste system," Team Six's colorful founder, Richard Marcinko, wrote in his autobiography, *Rogue Warrior,* "has the reputation of being about as rigid as any in the world." To get past those divisions, Marcinko broke ranks with strict naval protocols. He had the SEALs forgo standard dress codes and divisions between officers and enlisted: they wore what they wanted and rarely saluted each other. He also employed a time-tested bonding technique: getting drunk. Before deployment, he'd take his team out to a local Virginia Beach bar for one final bender. If there were any simmering tensions between members, they'd invariably come out after a few drinks. By morning, the men might be nursing headaches, but they'd be straight with each other and ready to function as a seamless unit.

Whether it's Marcinko's ad hoc methods for flipping the switch back in the eighties, or Davis's more refined approaches today, one critical issue remains: the ability to shut off the self and merge with the team is an exceptional and peculiar talent. That's why the SEALs have spent several decades developing such a rigorous filtration process. "If we really understood this phenomenon," says Davis, "we could train for it, not screen for it."

Unfortunately, screening is expensive and not that efficient. Nearly 80 percent of SEAL candidates wash out. They lose a ton of capable soldiers to the process. While it costs $500,000 to successfully train a SEAL, the cost of failure is tens of millions per year. Sure, some candidates fail to execute tactically—they shoot a cardboard hostage in the Kill House or drop a weapon out of a helicopter—but far more fail to synch up collectively. And this isn't surprising. Navigating ecstasis isn't in any field manual. It's a blank spot on their maps, beyond the pen of most cartographers, beyond the ken of rational folk.

But to the SEALs charged with capturing, not killing, Al-Wazu, it wasn't beyond the ken. It was just what happened out there. And, on that late September night, it happened quickly.

"The switch flipped as soon as we moved out," says Davis. "I could feel it, but I could also see it: the invisible mechanism locking in, the group synchronizing as we patrolled. The point man looking ahead, every man behind alternating their focus: one left, the next right, with rear security covering our six. Never walking backwards, but stopping, turning, scanning, then quickening the pace to catch up with the group, before doing it again. To look at it from a distance it would seem choreographed."

But it wasn't.

The patrol was quick. In less than twenty minutes they reached the compound: four buildings surrounded by a high concrete wall. They stopped for a moment, final checks, a slight reorganization, then lit out again in five groups of five. One group covered the west and north, another the east and south, a third stayed behind to watch their backs. The final two groups launched the main assault. Everyone knew his job. Silence was key. Radio calls were prohibited. "Talking is too slow," says Davis. "It complicates things."

The assault teams were over the wall and into the buildings, blazingly fast. The first room was empty, the second was crowded and dark. There were armed guards mixed in with unarmed women and children. Under these conditions, false positives are more the rule than the exception, and knowing when *not* to shoot becomes the difference between a successful mission and an international incident.

The conscious mind is a potent tool, but it's slow, and can manage only a small amount of information at once. The subconscious, meanwhile, is far more efficient. It can process more data in much shorter time frames. In ecstasis, the conscious mind takes a break, and the subconscious takes over. As this occurs, a number of performance-enhancing neurochemicals flood the system, including norepinephrine and dopamine. Both of these chemicals amplify focus, muscle reaction times, and pattern recognition. With the subconscious in charge and those neurochemicals in play, SEALs can read micro-expressions across dark rooms at high speeds.

So, when a team enters hostile terrain, they can break complex threats into manageable chunks. They quickly segment the battle space into familiar situations they know how to handle, like guards that need disarming or civilians that need corralling, and unfamiliar situations—a murky shape in a far corner—that may or may not be a threat. With their minds and movements tightly linked, the entire team executes simultaneously, chunking and disarming without hesitation or error.

That night in Afghanistan, there was no hesitation. The SEALs cleared those rooms in moments, left a couple of men behind to watch their prisoners, then moved into the next building. That was when they spotted him: Al-Wazu was there when they entered, sitting in a chair, an AK-47 slung over his shoulder.

Standard rules of engagement say an armed enemy is a dangerous enemy, but there was nothing standard about this situation. The man in front of them had escaped prison, trained other terrorists, and conducted brutal attacks. He had killed and, if given the chance, would again. But there was one small detail that every SEAL who entered the room had, in milliseconds, seen, processed, and acted upon—or, rather, not acted upon. The detail was that, at this particular moment, their target's eyes were closed. Wazu was fast asleep. It was a bloodless capture. None wounded, none killed. Absolutely perfect.

Of course, this isn't your typical war story. It's unlikely to make the news or get turned into a movie. Hollywood studios prefers lone heroes to faceless teams, and their accounts romanticize drama and disaster. But what the SEALs accomplished on that raid comes much closer to illustrating the true core of special operations culture: at their best, they are always an anonymous team. "I do not seek recognition for my actions . . . ," reads the SEAL code. "I expect to lead and be led . . . my teammates steady my resolve and silently guide my every deed." And this ethos is reinforced every time they flip that switch, when egos disappear and they perform together in ways that are just not possible alone.

The hardest part of a SEAL's job is knowing when *not* to shoot. Al-Wazu was hauled back to prison alive, and not one round had been fired. SEAL training is one of the most expensive filtration systems ever constructed, and it's largely designed to make ecstasis possible. So what's its real value?

"Well," says Davis, "when we shook Wazu awake, and he saw a group of steely-eyed, black-faced Navy SEALs in his living room—the look on his face? Priceless."

Google Goes Fishing

In a high desert valley, on the other side of the world from the SEALs' Afghan hunting grounds, Larry Page and Sergey Brin, the young founders of Google, realized they needed a better filter for "ecstasis" themselves.

And fast.

It was 2001, three years before Al-Wazu's rude awakening, and Page and Brin faced the biggest personnel decision of their start-up lives. Despite creating one of Silicon Valley's more notorious hiring gauntlets, where candidates were ruthlessly vetted for GPAs, SATs, and their ability to calculate MENSA-like brainteasers, the founders realized they couldn't crack this next hire with metrics alone.

After several years of rocket ship success, Google's board had decided that the company was growing too big for Larry's and Sergey's twenty-something britches. The investors felt a little "adult supervision" was needed and initiated a search for what would prove to be one of the more pivotal CEO hires of the high-tech era.

The process wasn't easy, on anyone. After nearly a year of interviews, as Brin later told the press, "Larry and I [had] managed to alienate fifty of the top executives in Silicon Valley." Time was running out. If they couldn't get it right soon, they'd prove the board's point: they were in over their heads.

In choosing their CEO, Page and Brin came to the conclusion that they had to look beyond their normal screening process. Resumes were all but useless. The technical part was more or less a given—there were plenty of sharp guys in the Valley who could run a stable of code monkeys. But, in a town full of outsize personalities, they had to find someone who could set ego aside and get what Google was trying to do. Someone who could, in the *New York Times*' John Markoff's assessment, "discipline Google's flamboyant, self-indulgent culture, without wringing out the genius."

Get it right, and they'd own the search engine space for a decade or more. Screw it up, and they could lose control of their company. Game over. Back to grad school.

So, in a stroke of desperate inspiration, Page and Brin found themselves turning to an unusual selection process, a brutal filtration system both strikingly similar to BUD/S and as wildly different as it could get.

Like the SEALs' infamous Hell Week, a finalist for Google's CEO job would have to spend five nearly sleepless days and nights enduring oppressive sun, freezing cold, and a 24/7 barrage of VUCA conditions. Pushed to physical and psychological extremes, the prospective leader would have nowhere to hide. Would he retreat into himself? Or could he merge with the team?

Of course, there were a few differences. Unlike the San Diego beach where BUD/S prospects prove themselves, the beach Page and Brin had in mind hadn't seen flowing water in nearly fifteen thousand years. It was now a bone-dry lake bed in the middle of Nevada's Black Rock Mountains. The site of Burning Man, one of the stranger rites of passage in modern times.

And rite of passage is the right phrase. This teeming, temporary carnival of tens of thousands has its own quirky customs, exotic rituals, and a fiercely dedicated following. It's a modern-day Eleusis, a Bacchanalian blowout, the Party at the End of Time—take your pick. But there's no denying the truth: something happens out there.

And Page and Brin were regular and enthusiastic attendees. The company that set the bar for catered perks ran free shuttle buses to the event. For many years, the two-story atrium of Building 43, Google's main headquarters, wasn't decorated with industry accolades or stock-ticker flat screens. Instead, it showcased pictures of loincloth-wearing, fire-spinning Googlers and their eclectic Burning Man art projects.

In fact, the very first Google Doodle, posted in the late summer of 1998, was a crude stick figure of the Burning Man himself.

Made from two commas set back to back, centered over the second yellow *"o"* in "*Google,*" that cryptic icon signified to those in the know that Page and Brin were turning out the lights in Palo Alto and lighting out for the Nevada badlands, uptime be damned.

So, when the founders heard that Eric Schmidt, the forty-six-year-old veteran of Sun Microsystems and a Berkeley Ph.D. computer scientist, was the sole CEO finalist who had already been to the event, they rejiggered their rankings and gave the guy a callback. "Eric was . . . the only one who went to Burning Man," Brin told Doc Searls, then a Berkman Center fellow at Harvard. "We thought [that] was an important criterion."

Stanford sociologist Fred Turner agrees, arguing that the festival's appeal to Silicon Valley is that it brings that hive mind experience to the masses. "[It] transforms the work of engineering into . . . a kind of communal vocational ecstasy." One of Turner's research subjects, a Googler himself, explained his experience on a pyrotechnic team: "[We were] very focused, very few words, open to anything . . . no egos. We worked very tightly. . . . I loved the 'feeling of flow' on the team—it was an extended, ecstatic feeling of interpersonal unity and timelessness we shared throughout."

And like the SEALs flipping the switch, the Googler's "communal vocational ecstasy" relies on changes in brain function. "Attending festivals like Burning Man," explains Oxford professor of neuropsychology Molly Crockett, "practicing meditation, being in flow, or taking psychedelic drugs rely on shared neural substrates. What many of these routes have in common is activation of the serotonin system."

But it's not only serotonin that makes up the foundation of those collaborative experiences. In those states, all of the neurochemicals that can arise—serotonin, dopamine, norepinephrine, endorphins, anandamide, and oxytocin—play roles in social bonding. Norepinephrine and dopamine typically underpin "romantic love," endorphins and oxytocin link mother to child and friend to friend, anandamide and serotonin deepen feelings of trust, openness, and intimacy. When combinations of these chemicals flow through groups at once, you get tighter bonds and heightened cooperation.

That heightened cooperation, that communal vocational ecstasy, was what Page, Brin, and so many of Google's engineers had discovered in the desert. It was an altered state of consciousness that suggested a better way of working together, and a feeling that anyone who presumed to lead them simply had to know firsthand. Maybe, if Schmidt could endure the blistering heat, the dust storms, the sleepless nights, and the relentless don't-give-a-shit-who-you-are strangeness of Burning Man, just maybe, he'd be the guy who could help them grow the dream without killing it.

Did it work? Did a bash in the boonies filter for critical talent better than any algorithm they could code? "The whole point of taking Schmidt to Burning Man," explains Salim Ismail, global ambassador for Singularity University and a Silicon Valley fixture, "was to see how he could handle a wild environment. Could he deal with the volatile, novel context? The extreme creativity? Did he merge with his team or stand in their way? And that's what they learned on that trip, that's one of Schmidt's great talents. He's really flexible, even in difficult conditions. He adapted his management style to fit their culture without bleeding out their genius and turned Google into a monster success."

Just check the numbers. When Google hired Schmidt in 2001, their revenues were rumored to be about $100 million. A decade later, when Schmidt finally handed the CEO reins back to Page, the company's revenues were nearly $40 billion.

That's a return of almost 40,000 percent.

Page and Brin have gone on to become numbers nine and ten on *Forbes*'s list of the world's wealthiest individuals, while Schmidt is one of the only nonfounder, non-family-members to ever become a stock option billionaire in history. Even for a company like Google, dedicated to unassuming goals like "10x moonshots" and organizing the entire world's information—a 400x return?

As close to priceless as they'll ever get.

Hacking Ecstasis

What's really going on here? Why did Google and the Navy SEALs, two of the highest-performing organizations in the world, have to resort to makeshift filters to find the next-level skills they desperately needed? After all, Page and Brin were two of the smarter Ph.D. students to come through Stanford in years. The team they gathered at Google was handpicked for its ability to quantify the inscrutable. Even back in 2001, the company was awash in cash. If there was a way to build or buy a better talent mousetrap, they would have used it to find their next CEO.

DEVGRU, meanwhile, has a blank check to pursue the cutting edge. In ammunition alone, annually, these guys spend as much as the entire U.S. Marine Corps. So for them to acknowledge, as Commander Rich Davis did, that an altered state of consciousness was both essential to mission success and elusive as hell—something they had to screen for by attrition, but couldn't train for by design? That doesn't make a lot of sense.

That's because, any way you slice it, ecstasis *doesn't* make a lot of sense. It remains a profound experience, a place far beyond our normal selves, what author Arthur C. Clarke called a "sufficiently advanced technology"—the kind that still looks like magic to us.

In light of this, it's easy to see why Google built their talent map around the reliable and observable: grade point averages, standardized tests, and IQ scores. It's what engineers know; it's how they think. SEALs, too, are famously empirical. If it doesn't work the first time, every time, they find something better that will. And theirs is a macho culture where feelings get short shrift.

So a feeling like ecstasis? No one's going to touch that one. Not, at least, until DARPA builds an implant for it.

So, ten years ago, this is where we found Google and the SEALs: two high-performing organizations hunting an odd set of skills that neither of them could name or train. And it's not that they were looking in the wrong place—they were just a little ahead of the curve.

Over the past ten years, science and technology have come round that bend. Empirical evidence has started to replace trial and error. And this is giving us new ways to approach ecstasis. But, before we dive into some of those stories, we first need to define our terms.

When we say *ecstasis* we're talking about a very specific range of nonordinary states of consciousness (NOSC)—what Johns Hopkins psychiatrist Stanislav Grof defined as those experiences "characterized by dramatic perceptual changes, intense and often unusual emotions, profound alterations in the thought processes and behavior, [brought about] by a variety of psychosomatic manifestations, rang[ing] from profound terror to ecstatic rapture . . . There exist many different forms of NOSC; they can be induced by a variety of different techniques or occur spontaneously, in the middle of everyday life."

Out of this broader inventory, we focused on three specific categories. First, flow states, those "in-the-zone" moments including group flow, or what the SEALs experienced during the capture of Al-Wazu, and the Googlers harnessed in the desert. Second, contemplative and mystical states, where techniques like chanting, dance, meditation, sexuality, and, most recently, wearable technologies are used to shut off the self. Finally, psychedelic states, where the recent resurgence in sanctioned research is leading to some of the more intriguing pharmacological findings in several decades. Taken together, these three categories define our territory of ecstasis.

Admittedly, these three may seem like strange bedfellows. And for most of the past hundred years, we've treated them that way. Flow states have been typically associated with artists and athletes; contemplative and mystical states belonged to seekers and saints; and psychedelic states were mostly sampled by hippies and ravers. But over the past decade, thanks to advances in brain science, we've been able to pull back the curtain and discover that these seemingly unrelated phenomena share remarkable neurobiological similarities.

Regular waking consciousness has a predictable and consistent signature in the brain: widespread activity in the prefrontal cortex, brainwaves in the high-frequency beta range, and the steady drip, drip of stress chemicals like norepinephrine and cortisol. During the states we're describing, this signature shifts markedly. Instead of widespread activity in the prefrontal cortex, we see specific parts of this region either light up and become *hyperactive* or power down and become *hypoactive.*

At the same time, brainwaves slow from agitated beta to daydreamy alpha and deeper theta. Neurochemically, stress chemicals like norepinephrine and cortisol are replaced by performance-enhancing, pleasure-producing compounds such as dopamine, endorphins, anandamide, serotonin, and oxytocin.

So no matter how varied these states appear on the surface, their underlying neurobiological mechanisms—that is, the *knobs and levers* being tweaked in the brain—are the same (see the endnotes for a thorough description). And this understanding allows us to tune altered states with newfound precision.

Consider one of the simplest and oldest ecstatic techniques: meditation. Historically, if you wanted to use meditation to consistently produce a state where the self vanished, decades of practice were required. Why? Because your target was nothing more than a peculiar sensation, and hitting it was like throwing darts blindfolded. But researchers now know that the center of that target actually correlates to changes in brain function—like brainwaves in the low-alpha, high-theta range—and this unlocks all kinds of new training options.

Instead of following the breath (or chanting a mantra or puzzling out a koan), meditators can be hooked up to neurofeedback devices that steer the brain directly toward that alpha/theta range. It's a fairly straightforward adjustment to electrical activity, but it can accelerate learning, letting practitioners achieve in months what used to take years.

For organizations like the SEALs and Google, these developments are allowing them to take an entirely different approach to high performance. They've moved beyond their earlier explorations, and are now pursuing ecstasis with a degree of precision that was simply not possible even ten years ago.

The Mind Gym

In the summer of 2013, we got a chance to meet with both the SEALs and Google, and see for ourselves how far they've come. We visited the SEALs because Rich Davis and several of DEVGRU's team leaders had read *The Rise of Superman* and noticed a considerable overlap between the flow described in the book and their own experiences on the battlefield. For Davis, that Al-Wazu raid was only one of dozens of missions where he'd found himself in the zone, doing the impossible. These moments changed his life. He began hunting for experts who could tell him how these states worked and how to get more of them. And while we were uncertain that we would have anything new to teach these guys, we got an invitation to the SEALs' Norfolk, Virginia, headquarters to observe the men in action and offer any insights we had on "flipping the switch."

After wading through several layers of background checks and byzantine paperwork, we spent a morning presenting to the teams and a few hours watching live-fire, hostage rescue training from an observation deck in the rafters of the Kill House. Then, during the debrief, we found ourselves sitting in a windowless conference room talking to team leaders about the high cost of screening for ecstasis. The issue wasn't just financial—the $500,000 it took to train a SEAL, the $4.25 million it cost to get them to DEVGRU, even the tens of millions wasted along the way; what concerned them more was the human cost. Again and again, we heard how emotionally devastating their screening process can be. How failure ruins careers and lives. "We're a very high-performing club," explained one SEAL team leader, and "some guys can't bounce back from failure."

When that meeting was over, they walked us through their newest facility, the Mind Gym, which was their best guess at how to train for ecstasis and not just screen for it. Sure, it cost millions to build, but if it could help them flip that switch reliably—if it could help more good men learn this invisible skill—it would be worth much more than that.

Equal parts CrossFit sweat and DARPA wizardry, the Mind Gym is a collection of some of the best tools and tech for training body-brain performance in the world: EEG brain monitors, medical-grade cardiac coherence devices, motion-tracking fitness stations, all kitted out with sensors, scanners, and screens designed to drive the SEALs into the zone faster than ever.

As we rounded one corner in the facility, we spotted four egg-shaped pods in a small alcove. They were sensory deprivation tanks, where users float in salt water in pitch blackness for hours at a time. Invented by National Institutes of Health researcher and neuroscientist John Lilly in the 1960s, these tanks were specifically designed to help people shut off the self (since the brain uses sensory inputs to help create our sense of self, by removing those inputs, you can dial down this sense). After Lilly began using these tanks to explore the effects of LSD and ketamine on consciousness, they fell out of favor with the establishment and devolved into a countercultural curiosity. But here they were again, in the red-hot center of the military-industrial complex, being used to train supersoldiers.

And the SEALs have been iterating on Lilly's original technology. Working with researchers at Advanced Brain Monitoring, in Carlsbad, California, they've hotwired neural and cardiac feedback loops, digital displays, and high-fidelity sound into the experience. They're deploying these upgrades for a practical purpose: accelerated learning. By using the tanks to eliminate all distraction, entrain specific brainwaves, and regulate heart rate frequency, the SEALs are able to cut the time it takes to learn a foreign language from six months to six weeks. For a specialized unit deployed across five continents, shutting off the self to accelerate learning has become a strategic imperative.

It's not just the Navy that is studying this domain in more depth. A few months after our visit to Norfolk, we crossed the country for a trip to the Googleplex. We were there to talk flow states with engineers, and learn more about what the company is doing to harness the "communal vocational ecstasy" they'd first glimpsed at Burning Man in the Black Rock Desert of Nevada.

Right after our presentation, we pedaled a couple of the ubiquitous and colorful Google bikes to the other side of campus to attend the opening of their new multimillion-dollar mindfulness center. Outfitted in soothing lime green with bamboo accents, the center features a vitality bar offering fresh-squeezed juices around the clock and a suite of meditation rooms decked out with sensor suits and neurofeedback devices similar to what we saw in the Navy's Mind Gym. Google had realized that when it comes to the highly competitive tech marketplace, helping engineers get into the zone and stay there longer was an essential investment. But like the SEALs, they hadn't completely ironed out all the variables.

"It's going well," explained Adam Leonard, one of the leaders of "G Pause" (their name for their mindfulness training program). "We've got active communities around the world, but the bigger challenge is getting people who aren't already meditators to start. The folks that already sit [in meditation] understand the benefits. It's the ones that are too busy and too stressed to slow down and need it the most that are the hardest to reach."

Not for lack of trying, though. In talking to Google's human performance team, we learned that many of the company's legendary efforts to create a seamless live/work environment—from Wi-Fi enabled commuter shuttles to farm-to-table dining rooms to pre-booked tickets for weekend adventures—were also attempts to minimize interruptions and keep employees in flow.

"Unlike those of many other firms," Stanford's Fred Turner points out, "Google's managers have subsidized the . . . explorations of its engineers and administrators and have promulgated relentlessly an ethos of benevolent peer production." By doing everything possible to keep people out of their heads and absorbed in their projects, Google is trying to make that same vocational ecstasy they found in the desert a permanent part of their on-campus lives.

The Altered States Economy

After those visits, and seeing how much time and money these two organizations were willing to put into maximizing the benefits of altered states, we couldn't help but wonder about the rest of us. Was it possible that deliberately seeking ecstasis went beyond high-performing organizations? Did any of this matter to regular folks? And if so, how much?

"Tell me what you value and I might believe you," management guru Peter Drucker once said, "but show me your calendar and your bank statement, and I'll show you what you *really* value." So we decided to take Drucker's advice and follow the money.

First, we dubbed the amount of cash and coin people spend each year trying to get out of their heads the "Altered States Economy." And we didn't mean this metaphorically; we meant it literally.

"Getting out of our heads" requires a precise biological signature in the brain. Specifically, a slowdown in neuroelectrical activity, a deactivation of the network that supports self-consciousness, and the presence of at least a couple of the "big six" neurochemicals we mentioned earlier. If an experience produces this signature, then we could credibly include it in our tally.

With neurobiology as our filter, we were able to spot similarities between otherwise disparate experiences. By paying attention to a singular category—like flow states or contemplative states or psychedelic states—it would have been easy to miss the larger trend and deeper patterns. But, with the knobs and levers serving as a "Rosetta Stone" for nonordinary consciousness, we could decode commonalities and measure impact in ways that were simply impossible before. In other words, we could start to put some hard numbers around the Altered States Economy.

Now, to be clear, we are not implying that all of the categories we are about to consider reflect deliberate, healthy, or intentional approaches to cultivating ecstasis. In fact, many are the exact opposite: impulsive, destructive, and unintentional. But that very fact—that we are driven to pursue altered states often at a steep cost—underscores how large and sometimes hidden a role they play in our lives.

We began our tally with the fairly uncontroversial assumption that any accounting of ecstasis should include all the substances people use to change states, from alcohol, tobacco, and caffeine on the licit side to cocaine, heroin, and methamphetamines on the illicit side (and if you're not sure that coffee should qualify as a state-changing drug, just look at the Starbucks line at 7 A.M.). We also included the legal and illegal markets for marijuana, psychopharmaceuticals like Ritalin and Adderall, and mood-shifting painkillers like OxyContin and Vicodin.

Next, we widened the net beyond substances that change our state of mind to experiences that do the same. We assessed therapeutic and personal development programs designed to "get me out of my head and help me feel happier," from psychological and psychiatric counseling to the massive online self-help market.

We also considered a wide range of high-flow pursuits like action sports, video games, and gambling—that is, activities that are primarily engaged for intrinsic reward, rather than external recognition.

Then we took a conservative approach to the broader categories of media and entertainment. While one could argue, for example, that much of the live music industry reflects a desire for state-changing collective experience, we zeroed in on an ascendant and uniquely qualified genre: electronic dance music (EDM). In EDM, leading DJs earn eight figures a year for showing up in a club and pressing "play" on a laptop. So it's not about the appeal of the band. There isn't one. And it's not about the lyrics, either. There aren't any. What is it about? Thunderous bass, tightly synchronized light shows, and, typically, lots of mind-altering substances. Other than the state-shift it produces, there is little reason to seek out the experience. And those states have become increasingly popular. In 2014, EDM represented almost *half* of all concert sales, attracting a quarter of a million concertgoers at a time and drawing the attention of Wall Street investors and major private equity firms.

We were equally focused in our assessment of film and TV, narrowing our accounting to genres that are especially immersive and escapist, like IMAX/3D films and streaming pornography. In the case of IMAX, for instance, why go to see these movies at all? In a few months, we could catch the identical film in the comfort of our homes. Instead, we drive to faraway theaters and pay a premium for total immersion: surround sound that shakes our seats, forty-foot screens that swallow our vision, and the company of others who gasp, boo, and clap alongside us. We don't pay extra to see more, we pay it to *feel* more—and think less.

And then there's pornography. Given that seven of the top twenty most-visited sites on the Web are porn sites, and that nearly

33 percent of all internet searches are for terms related to sex, it's safe to say that we're sinking a ton of time and money into digital voyeurism. Unlike analog sex, viewing porn has no evolutionary payoff. So why do so many do it so often? Because, for a brief moment (and it really is brief—an average PornHub visit clocks in at seven and half minutes), we lose ourselves in a state of physiological arousal and neurochemical saturation. Put bluntly, we watch porn to get high, not to get laid.

We ended our study with what many of us know best these days: social media. What makes these online distractions so sticky is how effectively they prime our brains for reward (mainly the feel-good neurochemical dopamine). Stanford neuroscientist Robert Sapolsky calls this priming the "magic of maybe." When we check our email or Facebook or Twitter, and sometimes we find a response and sometimes we don't, the next time a friend connects, Sapolsky discovered that we enjoy a 400 percent spike in dopamine. This can become distracting to the point of addicting. In 2016, the business consultancy Deloitte found that Americans are looking at their phones more than *eight billion* times a day. In a world where 67 percent of us admit to checking our status updates in the middle of the night, during sex, and before attending to basic biological needs like going to the bathroom, sleeping, or eating breakfast, we think it's safe to assume that a good part of what we're habitually doing online is more to forget ourselves for a moment than inform ourselves for the long haul.

Category by category, we followed Drucker's advice, seeing what our calendars and our bank accounts said about how much we really value stepping outside ourselves. And what we found was staggering. (See endnotes for a detailed workup of these numbers and www.stealingfirebook.com/downloads/ for a worksheet where you can calculate your own personal tally.)

Added all together, the Altered States Economy totals out to roughly $4 *trillion* a year. That's a sizable chunk of our income that we annually tithe to the Church of the Ecstatic. We spend more on this than we do on maternity care, humanitarian aid, and K–12 education *combined*." It's larger than the gross national product of Britain, India, or Russia. And to really put this in perspective, it's twice as many dollars as there are known galaxies in the entire universe. So even though much of our seeking is haphazard and often counterproductive, this $4 trillion total stands as a pretty good metric for how badly we want to get out of our heads, and how much we're willing to spend for even a shot at relief.

Yet this raises a few additional questions. If we're already spending a ton of time and money chasing these states, and even elite organizations like the SEALs and Google haven't definitively cracked the code, could something so elusive and confounding be worth all that trouble? Can these experiences provide benefits we can't get any other way? Put simply, are they worth it?

CHAPTER TWO

WHY IT MATTERS

The Ambassador of Ecstasis

In 2011, an out-of-work television host named Jason Silva posted a short, strange video on the internet. Titled "You Are a Receiver," the video was a two-minute barrage of quick-cut sci-fi imagery interspersed with shots of Silva, wearing jeans and a T-shirt, talking directly to the camera. What he was talking about was existential philosophy, evolutionary cosmology, and altered states of consciousness—that is, topics that don't usually show up in viral videos. In 2011, the Web's hottest fare were cartoon cats and honey badgers. But Silva's video struck a nerve, grabbing nearly half a million views in less than a month.

More videos followed. Between 2011 and 2015, Silva put more than a hundred different offerings online, garnering over 70 million views. NASA and *Time* reposted his work. The *Atlantic* ran a long profile, anointing him "the Timothy Leary of the Viral Video Age." Then the National Geographic Channel hired him to host "*Brain Games*," which became their highest-rated TV show ever and earned him an Emmy nomination. Yet, to Silva, all this attention came as something of a surprise: "When I started making videos, the goal wasn't celebrity. It was sanity."

Silva was born in Caracas, Venezuela, in 1982 and grew up during a turbulent time in the country's history. While raised in a middle-class family, his parents divorced when he was twelve and his father lost all his money when the Venezuelan economy collapsed in the late 1980s. There was an unsuccessful coup in 1992 and a successful coup in 2000. Crime and corruption skyrocketed. "Every member of my family was held up at gunpoint," recalls Silva. "My mother, my brother, even my grandmother. My father was kidnapped. I was a target. It was terrifying. It colored everything—my mom's not home by five P.M., so did she get kidnapped? Did she get killed? It was this constant, gnawing fear that never went away."

That fear turned Silva into a shut-in. By the time he was a teenager he could barely leave his house. He became paranoid, constantly wondering if all the doors were locked, if the noise he just heard was an intruder. "I was a kid," he says; "it was supposed to be this carefree time. But I was always battling crazy, neurotic thoughts and it was just crippling."

In high school, in an effort to recover sanity and a social life, Silva started organizing little gatherings at his house. "I was inspired by Baudelaire's hashish salons," he says. "So every Friday night a bunch of us would get together. Some people drank wine, some people smoked pot, but everyone talked philosophy. And those conversations would swallow me whole. I'd go off on a monologue and disappear. Totally out of my head. And it was exactly what I was searching for, a way to shut off my neurotic brain."

Quickly, Silva found these Friday nights shaping the rest of his week, as if those altered hours were overwriting those fearful years. He discovered a new sense of confidence. "I was always looking for my niche. I wasn't a great athlete, or the best student, or one of the cool kids. But those states showed me a part of myself I never knew existed. It started to feel like I had a superpower."

That's where the videos came in. At first, to ensure he wasn't just babbling, Silva had his friends record him during his rants. Later, he watched the tapes. "I was stunned. The stuff coming out of my mouth? Jaw-dropping connections between ideas. I had no idea where the insights were coming from. It was me, but it wasn't me."

And those videos led to film school in Miami, where he made even more videos. These efforts soon garnered attention. Because they saw his work and liked his screen presence, former vice president Al Gore's network, Current TV, hired him as a host. But it was a job he couldn't keep. "Current was great," he explains, "but most of what I did was read pop culture stories from a teleprompter. I didn't get to go off on crazy soliloquies, which meant I was cut off from flow. All that neurosis came flooding back. What I realized at Current was that I couldn't live without frequent access to these states. So I quit, and started making videos about them."

In Silva, ecstasis had found an ambassador. Because the conditions of his life and the wiring of his mind made his interior reality so uncomfortable, he got very good at tinkering with his consciousness. In his intuitive pursuit of these moments, Silva cobbled together a remarkably effective way to get outside himself for relief and inspiration. In high school, these states gave him back his life; in adulthood they gave him a career. "Really," he says, "what I found in altered states was freedom. First they gave me freedom from myself; then they gave me freedom to express myself, then they showed me what was actually possible. But it's not just me. I think almost every successful person I've met—one way or another—has found a way to use these states to propel them to levels they didn't know were possible."

And in saying "one way or another," Silva's getting at an important point. While the ways people get into these states vary considerably, their lived experiences share remarkable overlap. In fact, a big part of Silva's appeal hinges on this overlap. "A Buddhist monk experiencing satori while meditating in a cave, or a nuclear physicist having a breakthrough insight in the lab, or a fire spinner at Burning Man," he says, "look like different experiences from the outside, but they feel similar from the inside. It's a shared commonality, a bond linking all of us together. The ecstatic is a language without words that we all speak."

So, in the same way that the biological mechanisms underpinning certain non-ordinary states are remarkably consistent, our experiences of these states are, too. To be sure, the actual content will vary wildly across cultures: a Silicon Valley computer coder may experience a midnight epiphany as being in "the zone" and see streaming zeros and ones like the code from *The Matrix;* a French peasant girl might experience divine inspiration and hear the voice of an angel; an Indian farmer might see a vision of Ganesh in a rice paddy. But once we get past the narrative wrapping paper—what researchers call the "phenomenological reporting"—we find four signature characteristics underneath: Selflessness, Timelessness, Effortlessness, and Richness, or STER for short.

Certainly, researchers have come up with plenty of other descriptions of altered states, but we chose the four categories of STER for a specific reason. In reviewing the literature, we discovered that almost every previous breakdown of these experiences was weighed down by content. Trying to tease apart the consciousness-altering effects of meditation, for example, means wading through religious interpretations of what those states mean. Examine the academic criteria for flow and you'll find empirical triggers for how to produce the state mixed in with the subjective experiences of the state. The same goes for many of the psychedelic rating scales, which often presuppose that future subjects will have a similar range of experiences (ranging from nature mysticism, to natal regression, to cosmic union) as the original experimenters.

But the four categories we've zeroed in on are content neutral. They're a strictly phenomenological description (how these states make us feel) rooted in shared neurobiology. This gets us past initial preconceptions about what these experiences are supposed to mean or reveal. While there's still much work to be done, we've now introduced this model to researchers from Harvard, Stanford, Yale, and Oxford, and they've found it useful. It's experimental and experiential and we hope it can help simplify and integrate the ongoing conversation around altered states. (And if you're interested in helping further this research, visit: www.stealingfreebook.com/research/.)

Selflessness

Despite all the recent talk about supercomputers and artificial intelligence, the human brain remains the most complex machine on the planet. At the center of this complexity lies the prefrontal cortex, our most sophisticated piece of neuronal hardware. With this relatively recent evolutionary adaptation came a heightened degree of self-awareness, an ability to delay gratification, plan for the long term, reason through complex logic, and think about our thinking. This hopped-up cogitation promoted us from slow, weak, hairless apes into tool-wielding apex predators, turning a life that was once nasty, brutish, and short into something decidedly more civilized.

But all of this ingenuity came at a cost. No one built an off switch for the potent self-awareness that made it all possible. "[T]he self "is not an unmitigated blessing," writes Duke University psychologist Mark Leary in his aptly titled book, *The Curse of the Self.* "It is single-handedly responsible for many, if not most of the problems that human beings face as individuals and as a species . . . [and] conjures up a great deal of personal suffering in the form of depression, anxiety, anger, jealousy, and other negative emotions." When you think about the billion-dollar industries that underpin the Altered States Economy, isn't this what they're built for? To shut off the self. To give us a few moments of relief from the voice in our heads.

So, when we do experience a non-ordinary state that gives us access to something more, we feel it first as something *less*—and that something missing is us. Or, more specifically, the inner critic we all come with: our inner Woody Allen, that nagging, defeatist, always-on voice in our heads. You're too fat. Too skinny. Too smart to be working this job. Too scared to do anything about it. A relentless drumbeat that rings in our ears.

This was Silva's monologue too, but he stumbled onto a curious fact—altered states can silence the nag. They act as an off switch. In these states, we're no longer trapped by our neurotic selves because the prefrontal cortex, the very part of the brain generating that self, is no longer open for business.

Scientists call this shutdown "transient hypofrontality." *Transient* means temporary. "*Hypo,*" the opposite of "*hyper,*" means "less than normal." And *frontality* refers to the prefrontal cortex, the part of our brain that generates our sense of self. During transient hypofrontality, because large swatches of the prefrontal cortex turn off, that inner critic comes offline. Woody goes quiet.

Without all the badgering, we get a real sense of peace. "This peacefulness may result from the fact," continues Leary, that "without self-talk to stir up negative emotions, the mystical experience is free of tension." And with tension out of the way, we often discover a better version of ourselves, more confident and clear.

"For me," explains Silva, "it's a simple equation. If I hadn't learned to shut off the self, I'd be the same mess I was back in Venezuela. Too fearful to do much of anything. But once the voice in my head disappears, I get out of my own way."

And the benefits of selflessness go beyond silencing our inner critic. When free from the confines of our normal identity, we are able to look at life, and the often repetitive stories we tell about it, with fresh eyes. Come Monday morning, we may still clamber back into the monkey suits of our everyday roles—parent, spouse, employee, boss, neighbor—but, by then, we know they're just costumes with zippers.

Psychologist Robert Kegan, chair of adult development at Harvard, has a term for unzipping those costumes. He calls it "the subject-object shift" and argues that it's the single most important move we can make to accelerate personal growth. For Kegan, our *subjective* selves are, quite simply, who we think we are. On the other hand, the "objects" are things we can look at, name, and talk about with some degree of *objective* distance. And when we can move from being subject to our identity to having some objective distance from it, we gain flexibility in how we respond to life and its challenges.

In time, Silva noticed exactly this change. "Whenever I get out of my head, I get a little more perspective. And every time I return, my world is a little bit wider and I'm a little bit less neurotic. Over the years, it's made a real difference."

That's Kegan's point. When we are reliably able to make the subject-object shift, as he points out in his book *In Over Our Heads,* "You start . . . constructing a world that is much more friendly to contradiction, to oppositeness, to being able to hold onto multiple systems of thinking. . . . This means that the self is more about movement through different forms of consciousness than about defending and identifying with any one form."

By stepping outside ourselves, we gain perspective. We become objectively aware of our costumes rather than subjectively fused with them. We realize we can take them off, discard those that are worn out or no longer fit, and even create new ones. That's the paradox of selflessness—by periodically losing our minds we stand a better chance of finding ourselves.

Timelessness

A quick search on Google yields over 11.5 billion hits for the word "*time.*" In comparison, more obvious topics of interest like sex and money rank a paltry 2.75 billion and 2 billion, respectively. Time, and how to make the most of it, appears to be about five times more important to us than making love or money.

And there's good reason for this obsession. According to a 2015 Gallup survey, 48 percent of working adults feel rushed for time, and 52 percent report significant stress as a result. Bosses, colleagues, kids, and spouses all expect instant response to emails and texts. We never really get free of our digital leashes, even in bed or on vacation. Americans are now working longer hours with less vacations than any industrialized country in the world.

"Time poverty," as this shortage is known, comes with consequences. "When [you] are juggling time," Harvard economist Sendhil Mullainathan recently told the *New York Times*, ". . . you borrow from tomorrow, and tomorrow you have less time than you have today. . . . It's a very costly loan."

Non-ordinary states provide some relief from this rising debt, and they do it in much the same way as they quiet our inner critic. Our sense of time isn't localized in the brain. It's not like vision, which is the sole responsibility of the occipital lobes. Instead, time is a distributed perception, calculated all over the brain, calculated, more specifically, all over the prefrontal cortex. During transient hypofrontality, when the prefrontal cortex goes offline, we can no longer perform this calculation.

Without the ability to separate past from present from future, we're plunged into an elongated present, what researchers describe as "the deep now." Energy normally used for temporal processing gets reallocated for focus and attention. We take in more data per second, and process it more quickly. When we're processing more information faster, the moment seems to last longer—which explains why the "now" often elongates in altered states.

When our attention is focused on the present, we stop scanning yesterday for painful experiences we want to avoid repeating. We quit daydreaming about a tomorrow that's better than today. With our prefrontal cortex offline, we can't run those scenarios. We lose access to the most complex and neurotic part of our brains, and the most primitive and reactive part of our brains, the amygdala, the seat of that fight-or-flight response, calms down, too.

In his book *The Time Paradox,* Stanford psychologist Philip Zimbardo, one of the pioneers in the field of time perception, describes it this way: "When you are . . . fully aware of your surroundings and of yourself in the present, [this] increases the time that you swim with your head above water, when you can see both potential dangers and pleasures. . . . You are aware of your position and your destination. You can make corrections to your path."

In a recent study published in *Psychological Science*, Zimbardo's Stanford colleagues Jennifer Aaker and Melanie Rudd found that an experience of timelessness is so powerful it shapes behavior. In a series of experiments, subjects who tasted even a brief moment of timelessness "felt they had more time available, were less impatient, more willing to volunteer to help others, more strongly preferred experiences over material products, and experienced a greater boost in life satisfaction."

And when we do slow life down, we find the present is the only place in the timescape we get reliable data anyway. Our memories of the past are unstable and constantly subject to revision—like a picture-book honeymoon overwritten by a bitter divorce. "[M]emory distortions are basic and widespread in humans," acknowledges cognitive psychologist Elizabeth Loftus, "and it may be unlikely that anyone is immune." The past is less an archived library of what really happened, and more a fluid director's commentary we're constantly updating.

Future forecasts aren't much better. When we try to predict what's around the bend, we rarely get it right. We tend to assume the near future will look much like the recent past. That's why events like the toppling of the Berlin Wall and the 2008 financial collapse caught so many analysts flatfooted. What looks inevitable in hindsight is often invisible with foresight.

But when non-ordinary states trigger timelessness, they deliver us to the perpetual present—where we have undistracted access to the most reliable data. We find ourselves at full strength. "That was another thing I noticed," says Silva, "when I go off on a tangent and the ideas start to flow, there's no room for anything else. Definitely not for time. People who see my videos often ask how I can find all those connections between ideas. But the reason I can find them is simple: without time in the picture, I have all the time I need."

Effortlessness

These days, we're drowning in information, but starving for motivation. Despite a chirpy self-improvement market peppering us with endless tips and tricks on how to live better, healthier, wealthier lives, we're struggling to put these techniques into action. One in three Americans, for example, is obese or morbidly obese, even though we have access to better nutrition at lower cost than at any time in history. Eight out of ten of us are disengaged or actively disengaged at work, despite the HR circus of incentive plans, team-building off-sites, and casual Fridays. Big-box health clubs oversell memberships by 400 percent in the certain knowledge that, other than the first two weeks in January and a brief blip before spring break, fewer than one in ten members will ever show up. And when a Harvard Medical School study confronted patients with lifestyle-related diseases that would kill them if they didn't alter their behavior (type 2 diabetes, smoking, atherosclerosis, etc.), 87 percent couldn't avoid this sentence. Turns out, we'd rather die than change.

But just as the selflessness of an altered state can quiet our inner critic, and the timelessness lets us pause our hectic lives, a sense of *effortlessness* can propel us past the limits of our normal motivation.

And we're beginning to understand where this added drive comes from. In flow, as in most of the states we're examining, six powerful neurotransmitters—norepinephrine, dopamine, endorphins, serotonin, anandamide, and oxytocin—come online in varying sequences and concentrations. They are all pleasure chemicals. In fact, they're the six most pleasurable chemicals the brain can produce and these states are one of the only times we get access to many of them at once. That's the biological underpinning of effortlessness: "I did it, it felt *awesome*, I'd like to do it again as soon as possible."

When psychologist Mihaly Csikszentmihalyi did his initial research into flow, his subjects frequently called the state "addictive," and admitted to going to exceptional lengths to get another fix. "The [experience] lifts the course of life to another level," he writes in his book *Flow*. "Alienation gives way to involvement, enjoyment replaces boredom, helplessness turns into a feeling of control. . . . When experience is intrinsically rewarding life is justified."

So, unlike the slog of our to-do lists, once an experience starts producing these neurochemicals, we don't need a calendar reminder or an accountability coach to make sure we keep doing it. The intrinsically rewarding nature of the experience compels us. "So many people find this so great and high an experience," wrote psychologist Abraham Maslow in his book *Religion, Values, and Peak Experiences*, "that it justifies not only itself, but living itself."

This explains why Silva "couldn't live without access to these states" and left a great job at Current TV for the uncertain prospect of making more videos. It's why action and adventure athletes routinely risk life and limb for their sports and why spiritual ascetics willingly trade creature comforts for a chance to glimpse God. "In a culture supposedly ruled by the pursuit of money, power, prestige, and pleasure," Csikszentmihalyi wrote in *Beyond Boredom and Anxiety*, "it is surprising to find certain people who sacrifice all those goals for no apparent reason. . . . By finding out why they are willing to give up material rewards for the elusive experience of performing enjoyable acts we . . . learn something that will allow us to make everyday life more meaningful."

But you don't have to take extreme risk or give up material reward to experience this benefit. It shows up wherever people are deeply committed to a compelling goal. When John Hagel, the cofounder of Deloitte consulting's Center for the Edge, made a global study of the world's most innovative, high-performing business teams—meaning the most motivated teams on the planet—he too found that "the individuals and organizations who went the farthest the fastest were always the ones tapping into passion and finding flow."

This ability to unlock motivation has widespread implications. Across the board, from education to health care to business, motivational gaps cost us trillions of dollars a year. We *know* better; we just can't seem to *do* better. But we can do better. Effortlessness upends the "suffer now, redemption later" of the Protestant work ethic and replaces it with a far more powerful and enjoyable drive.

Richness

The final characteristic of ecstasis is "richness," a reference to the vivid, detailed, and revealing nature of non-ordinary states. In his first video, "You Are a Receiver," Silva explains it like this: "It's creative inspiration or divine madness or that kind of connection to something larger than ourselves that makes us feel like we understand the intelligence that runs throughout the universe."

The Greeks called that sudden understanding *anamnesis.* Literally, "the forgetting of the forgetting." A powerful sense of remembering. Nineteenth century psychologist William James experienced this during his Harvard experiments with nitrous oxide and mescaline, noting it's "the extremely frequent phenomenon, that sudden feeling . . . which sometimes sweeps over us, having "been here before", as if at some indefinite past time, in just this place . . . we were already saying just these things." And that feeling, of waking up to some ineffable truth that's been in us all along, can feel deeply *significant.*

In non-ordinary states, the information we receive can be so novel and intense that it feels like it's coming from a source outside ourselves. But, by breaking down what's going on in the brain, we start to see that what feels supernatural might just be supernatural: beyond our normal experience, for sure, but not beyond our actual capabilities.

Often, an ecstatic experience begins when the brain releases norepinephrine and dopamine into our system. These neurochemicals raise heart rates, tighten focus, and help us sit up and pay attention. We notice more of what's going on around us, so information normally tuned out or ignored becomes more readily available. And besides simply increasing focus, these chemicals amp up the brain's pattern recognition abilities, helping us find new links between all this incoming information.

As these changes are taking place, our brainwaves slow from agitated beta to calmer alpha, shifting us into daydreaming mode: relaxed, alert, and able to flit from idea to idea without as much internal resistance. Then parts of the prefrontal cortex begin shutting down. We experience the selflessness, timelessness, and effortlessness of transient hypofrontality. This quiets the "already know that, move along" voice of our inner critic and dampens the distractions of the past and future. All these changes knock out filters we normally apply to incoming data, giving us access to a fresh perspectives and more potential combinations of ideas.

As we move even deeper into ecstasis, the brain can release endorphins and anandamide. They both decrease pain, removing the diversion of physical distress from the equation, letting us pay even more attention to what's going on. Anandamide also plays another important role here, boosting "lateral thinking," which is our ability to make far-flung connections between disparate ideas. Post-its, Slinkys, Silly Putty, Super Glue, and a host of other breakthroughs all came when an inventor made a sideways leap, applying an overlooked tool in a novel way. In part, that's anandamide at work.

And, if we go really deep, our brainwaves shift once again, pushing us toward quasi-hypnotic theta, a wave we normally produce only during REM sleep that enhances both relaxation and intuition. To wrap it all up, we can experience an afterglow of serotonin and oxytocin, prompting feelings of peace, well-being, trust, and sociability, as we start to integrate the information that has just been revealed.

And *revealed* is the right word. Conscious processing can only handle about 120 bits of information at once. This isn't much. Listening to another person speak can take almost 60 bits. If two people are talking, that's it. We've maxed out our bandwidth. But if we remember that our unconscious processing can handle billions of bits at once, we don't need to search outside ourselves to find a credible source for all that miraculous insight. We have terabytes of information available to us; we just can't tap into it in our normal state.

Umwelt is the technical term for the sliver of the data stream that we normally apprehend. It's the reality our senses can perceive. And all umwelts are not the same. Dogs hear whistles we cannot, sharks detect electromagnetic pulses, bees see ultraviolet light—while we remain oblivious. It's the same physical world, same bits and bytes, just different perception and processing. But the cascade of neurobiological change that occurs in a non-ordinary state lets us perceive and process more of what's going on around us and with greater accuracy. In these states, we get upstream of our umwelt. We get access to increased data, heightened perception, and amplified connection. And this lets us see ecstasis for what it actually is: an information technology. Big Data for our minds.

Wicked Solutions to Wicked Problems

Now that we've mapped out the biology and phenomenology beneath STER, we're going to turn our attention to a different couple of questions: While these states may make us feel better, can they help us think better? Do these short-term peaks enable us to solve real-world problems?

In 2013 we were invited to participate in the Red Bull Hacking Creativity project, a joint effort involving scientists at the MIT Media Lab, a group of TED Fellows, and the namesake energy drink company. Conceived by Dr. Andy Walshe, Red Bull's director of high performance (and a member of Flow Genome Project's advisory board), the project was the largest meta-analysis of creativity research ever conducted, reviewing more than thirty thousand research papers and interviewing hundreds of other subject-matter experts, from break dancers and circus performers to poets and rock stars. "It was an impossible goal," Walshe explained, "but I figured if we could crack something as hard to pin down as creativity, we could figure out almost anything after that."

As of late 2016, with the initial phases of the research completed, the study came to two overarching conclusions. First, creativity is essential for solving complex problems—the kinds we often face in a fast-paced world. Second, we have very little success training people to be more creative. And there's a pretty simple explanation for this failure: we're trying to train a skill, but what we really need to be training is a state of mind.

Conventional logic works really well for solving discrete problems with definite answers. But the "wicked problems" of today require more creative responses. These challenges defy singular stable solutions: issues as serious as war or poverty, or as banal as traffic and trends. Throw money, people, or time at any of these and you may fix a symptom, but you create additional problems: financial aid to the developing world, for example, often breeds corruption in addition to its intended relief; adding more lanes to the highway encourages more drivers and more gridlock; fighting wars to make the world safer can make it more dangerous than ever.

Solving wicked problems requires more than a direct assault on obvious symptoms. Roger Martin of the University of Toronto's Rotman School of Management conducted a lengthy study of exceptional leaders stretching from Procter & Gamble's then-CEO A. G. Lafley to choreographer Martha Graham and discovered that their ability to find solutions required holding conflicting perspectives and using that friction to synthesize a new idea. "The ability to face constructively the tension of opposing ideas," Martin writes in his book *The Opposable Mind,* ". . . is the only way to address this kind of complexity."

But developing Martin's "opposable mind" isn't easy. You have to give up exclusively identifying with your own, singular point of view. If you want to train this kind of creativity and problem solving, what the research shows is that the either/or logic of normal consciousness is simply the wrong tool for the job.

Scientists have discovered a better tool. The amplified information processing and perspective that non-ordinary states provide can help solve these types of complex problems, and they can often do so faster than more conventional approaches. Take meditation. Research done on Tibetan Buddhists in the 1990s showed that longtime contemplative practice can produce brainwaves in the gamma range. Gamma waves are unusual. They arise primarily during "binding," when novel ideas come together for the first time and carve new neural pathways. We experience binding as "Ah-Ha insight," that eureka moment, the telltale signature of sudden inspiration. This meant that meditation could amplify complex problem solving, but, since the monks needed to put in more than 34,000 hours (roughly thirty years) to develop this skill, it was a finding with limited application.

So researchers began to consider the impact of short-term meditation on mental performance. Was it possible, they wondered, to cut some monastic corners and still get similar results? Turns out, you can cut quite a few corners. Initial studies showed eight weeks of meditation training measurably sharpened focus and cognition. Later ones whittled that down to five weeks.

Then, in 2009, psychologists at the University of North Carolina found that even four days of meditation produced significant improvement in attention, memory, vigilance, creativity, and cognitive flexibility. "Simply stated," lead researcher Fadel Zeidan explained to *Science Daily,* "the profound improvements we found after just four days of meditation training are really surprising. . . . [They're] comparable to results that have been documented after far more extensive training." Rather than pulling a caffeinated all-nighter to force a eureka insight, or devoting decades to becoming a monk, we now know that even a few days' training in mindfulness can up the odds of a breakthrough considerably.

In the field of flow research, we see the same thing: being "in the zone" significantly boosts creativity. In a recent University of Sydney study, researchers relied on transcranial magnetic stimulation to induce flow—using a weak magnetic pulse to knock out the prefrontal cortex and create a twenty-to-forty-minute flow state. Subjects were then given a classic test of creative problem solving: the nine-dot problem. Connect nine dots with four lines without lifting pencil from paper in ten minutes. Under normal circumstances, fewer than 5 percent of the population pulls it off. In the control group, no one did. In the flow-induced group, 40 percent connected the dots in record time, or eight times better than the norm.

And this isn't a one-off finding. When neuroscientists at

DARPA and Advanced Brain Monitoring used a different technique—neurofeedback—to prompt flow, they found that soldiers solved complex problems and mastered new skills up to 490 percent faster than normal. It's for this reason that, when the global consultancy McKinsey did a ten-year global study of companies, they found that top executives—meaning those most called upon to solve strategically significant "wicked problems"—reported being up to 500 percent more productive in flow.

Similar results have also been showing up in psychedelic research. Several decades ago, James Fadiman, a researcher at the International Foundation for Advanced Study, in Menlo Park, California, helped bring together twenty-seven test subjects—mainly engineers, architects, and mathematicians drawn from places like Stanford and Hewlett-Packard—for one specific reason: for months prior, each of them had been struggling (and failing) to solve a highly technical problem.

Test subjects were divided into groups of four, with each group receiving two treatment sessions. Some were given 50 micrograms of LSD; others took 100 milligrams of mescaline. Both are microdosages, well below the level needed to produce psychedelic effects. Then subjects took tests designed to measure nine categories of cognitive performance enhancement (from heightened concentration to the ability to know when the right solution presents itself), and spent four hours working on their problems.

While everyone experienced a boost in creativity—some as much as 200 percent—what got the most attention were the real-world breakthroughs that emerged: "Design of a linear electron accelerator beam-steering device, a mathematical theorem regarding NOR-gate circuits, a new design for a vibratory microtome, a space probe designed to measure solar properties, and a new conceptual model of a photon."

None of these practical, technical achievements are the kind of result that most people associate with the navel-gazing world of psychedelics. But similar outcomes are happening in Fadiman's current survey of microdosing among professionals. With more than four hundred responses from people in dozens of fields, the majority, as Fadiman recently explained, report "enhanced pattern recognition [and] can see more of the pieces at once of a problem they are trying to solve."

With these developments, psychedelics have begun moving from recreational diversion to performance-enhancing supplement. "A shift began about four or five years ago," author and venture capitalist Tim Ferriss told us. "Once Steve Jobs and other successful people began recommending the use of psychedelics for enhancing creativity and problem solving, the public became a little more open to the possibility."

And, as Ferriss explained on CNN, it wasn't just the cofounder of Apple who made the leap. "The billionaires I know, almost without exception, use hallucinogens on a regular basis. These are people who are trying to be very disruptive. They look at problems in the world and they try to ask entirely new questions."

Wicked problems are those without easy answers—where our rational, binary logic breaks down and our normal tools fail us. But the information richness of a nonordinary state affords us perspective and allows us to make connections where none may have existed before. And it doesn't seem to matter which technique we deploy: mindfulness training, technological stimulation or pharmacological priming, the end results are substantial. Consider the gains: a 200 percent boost in creativity, a 490 percent boost in learning, a 500 percent boost in productivity.

Creativity, learning, and productivity are essential skills and those percentage gains are big numbers. If they were merely the result of a few studies done by a couple of labs, they would be easier to dismiss. But there is now seven decades of research, conducted by hundreds of scientists on thousands of participants, showing that when it comes to complex problem solving, ecstasis could be the "wicked solution" we've been looking for.

CHAPTER THREE

WHY WE MISSED IT

Beyond the Pale

In 1172, the English invaded Ireland, planted their flag, and built a great big fence. That barrier, known as the English Pale—from *pale,* meaning a stake or picket—defined the world for those invaders. Within their pale, all was safe, true, and good, a civilized land ruled by English law and institutions.

Beyond the pale, on the other hand, lay bad news. That's where mayhem, murder, and madness resided. Most who ventured beyond it were never heard from again. And the few who did manage to return weren't always welcomed with open arms. They were no longer trustworthy; they might have seen too much.

So, if you ask the question—where has this Promethean revolution been hiding—beyond the pale is a big part of the answer. That's because the experiences at the center of this book stand outside the perimeter fence of polite society. Instead of hearing stories about the possibilities of altered states, we're treated to cautionary tales. Stories of hubris and excess. Icarus redux.

This bias has obscured our view. It's clouded our judgment and cut us off from vital parts of ourselves and our potential. To get a better understanding of precisely how this has happened, we'll meet a jack Mormon rock star, a cyborg philosopher, and a disgraced scientist. Through their stories, we'll examine three instances of our current Pale—the Pale of the Church, the Pale of the Body, and the Pale of the State. We'll detail the historical reasons for each, and explore why understanding the role of the Pale is essential for anyone exploring ecstasis.

Let's start with the rock star.

The Pale of the Church

James Valentine is a tall, thin man in his late thirties, with shoulder-length straight brown hair, gray-blue eyes, and a scraggly beard. In person, he's thoughtful and soft-spoken. Onstage, as lead guitarist for Maroon 5, he's one of the more successful musicians in the world. Over the past fifteen years, there's rarely been a time when the band didn't have a song on the *Billboard* charts. They've won just about every music award, including three Grammys, three People's Choice Awards, and three MTV Music Awards. Yet, had a thirteen-year-old Valentine not bumped into the Holy Ghost while rounding first base, none of this would have happened.

That encounter took place in 1991, at a baseball field in Lincoln, Nebraska. Valentine, a devout member of the Church of Jesus Christ of Latter-day Saints, came from a religious family. His ancestors were Mormon pioneers, among those early faithful who fled religious persecution in Illinois and eventually settled their wagon trains in Salt Lake City, Utah. His grandfather was the mission president for South America; his aunt was secretary to the church's top leader, Prophet Thomas Monson. His father taught literature at Brigham Young University, while his brother and three sisters all graduated from the school.

This was supposed to be Valentine's path, too. He would finish high school, go on a mission—a two-year voluntary stint devoted to proselytizing and humanitarian aid work—and then return home to attend BYU and a life spent serving his church. Until that baseball game got in the way.

To understand what happened to Valentine during that game, we need to understand that Mormons believe the Holy Ghost can enter a person during prayer. "The feeling of spirit entering you," explains Valentine, "what Mormons call 'the feeling of the Holy Ghost,' is the very center of the religion. And it's a real sensation, a burning in the bosom that becomes a deeply joyful sense of peace and connection to something much greater than yourself."

Yet, there was absolutely no reason that Valentine should have rounded first base and bumped into the Holy Ghost. There is nothing particularly sacred about baseball. "It didn't make any sense," he explains. "I was a spiritual kid. I'd had plenty of experiences when the spirit entered me. But all of them took place in church, while praying. Not on a baseball field. It was incredibly confusing. I mean, as far as I knew, the Holy Ghost didn't really play baseball."

His confusion triggered a brief crisis of faith. But the real trouble showed up later that same year, when Valentine picked up a guitar. "When I started playing I also started having these crazy peak experiences," he explains. "Music was a direct pipeline into another world. And the feeling I got was exactly like the feeling of the Holy Ghost—the same feeling I bumped into rounding first base—only much more powerful. I would get sucked into these intense trances that would last for hours. I'd get so lost that drool would pour out of my mouth. I wouldn't even notice. And maybe the Holy Spirit was okay with baseball, but rock-'n'-roll? That was just totally out of bounds. But everything that's happened since, my whole career, has been an attempt to chase this feeling."

The pale that Valentine ventured beyond, call it the Pale of the Church, is an age-old barrier for the spiritually curious. It's a divide between those who believe that direct access to God should be moderated by a learned elite and those who believe direct access should be available to anyone at any time. Top-down versus bottom-up.

Ecstasy in small doses, metered out by those in charge, is a time-honored technique for social bonding and bureaucratic control. But ecstasy from a fire hose, filling anyone who asks with vibrant certainty, never mind the doctrine? That's downright dangerous.

In Christianity, it shows up as the tension between chapter-and-verse Roman Catholics and holy-rolling Pentecostals; in Islam, it's solemn imams versus twirling Sufis; in China, it's by-the-book Confucians against go-with-the-flow Taoists. In each case, a small community figures out a more direct path to knowledge and, because they blossom without the sanction of the orthodoxy, they are persecuted for it.

Spiritual subcultures that slip through heaven's gate tend to piss off the gatekeepers. In Valentine's case, once he realized the church wasn't his only access to the Holy Ghost, his dependence on organized religion dwindled. By the time he was sixteen, he told his father he wasn't going on a mission; by the time he was eighteen, he had left home for a life in rock and roll. But none of this was easy. "The Pale of the Church had a real hold on me," admits Valentine. "I was terrified to venture beyond it. I had no real idea what was going to happen."

Compared to many who came before him, Valentine got off lightly. Historically, we denounce these seekers in the strongest terms available. Consider Joan of Arc, the medieval French peasant girl who heard angels and led her nation to victory in the Hundred Years' War. She won battles, restored a king, and shook Europe's political, military, and religious structures to their core. But because she was an unordained woman claiming to know God's will, she found herself on the wrong side of the Pale of the Church, with tragic consequences.

At her trial for heresy, church officials denied her a lawyer, stacked the jury against her, and sent death threats to the judge. Then they tried to trap her with an impossible question: "Do you believe you experience the grace of God?"

If she answered "no," then she'd be admitting that the voices she heard were diabolical, not divine—and she would have to die for it. If she answered "yes," that she definitely knew she was in God's grace, then she'd have violated one of the core tenets of their doctrine—and she would have to die for it. Joan dodged both responses elegantly: "If I am not [in a state of grace], may God put me there; and if I am, may God so keep me."

She evaded the legalistic trap set for her, and at the same time affirmed her blameless faith. "Those who were interrogating her," the court notary remarked, "were stupefied."

But even that inspired testimony was not enough to save her young life. The presiding bishop replaced the nuns who should have been guarding her with soldiers who tried to rape her instead. Joan put men's breeches on to protect her honor, and then tied all of her leggings, hose, and tunic together in a kind of jury-rigged chastity belt.

The bishop seized the moment, condemning her for the lesser charge of cross-dressing-as-heresy. She had stolen fire and, the Church insisted, she'd die by it. They had her burned at the stake three times over, and so no one could gather a relic, dumped her ashes into the Seine.

The Pale of the Church is why, despite millennia of bold experimentation, the insights of mystics rarely survive. Their beliefs are ridiculed and their motives maligned. Lest anyone try to follow in their footsteps, their recipes for ecstasy are torn to pieces and scattered to the wind. Even when religions are built on the epiphanies of their founders, attempts to repeat those original experiments are strongly discouraged. It's one of the main reasons we've failed to notice the possibilities of non-ordinary states.

The Pale of the Church has been blocking our view.

The Pale of the Body

In the late 1990s, University of Edinburgh philosopher Andy Clark was researching cyborgs when he realized we were closer to that man-machine merger than anyone wanted to admit. If you have a pacemaker, cochlear implant, or even a pair of eyeglasses, you're using technology to upgrade biology. What Clark found strange was that no one seemed to notice these developments. Type "cyborg" into a search bar and the first thing that pops up is "a fictional or hypothetical person whose physical abilities are extended beyond normal human limitations by mechanical elements." But there's nothing fictional about eyeglasses or smartphones (or, for that matter, artificial hearts and bionic limbs). So why, Clark wanted to know, haven't we acknowledged that we're already becoming cyborgs?

What he realized is that we are confined by a cultural assumption—call it the Pale of the Body—that ranks the stuff we are (biology) above the stuff we make (technology). "[It is] the prejudice that whatever matters about my mind," Clark explains in his book *Natural-Born Cyborgs,* "depends solely on what goes on inside my biological skin-bag, inside the ancient fortress of skin and skull." We have trouble admitting we're cyborgs then, because the very prospect of augmenting ourselves with technology seems suspect.

And the skin-bag bias extends beyond tools that augment our bodies, to techniques that enhance our minds. In 2012, a study conducted by the American Pediatric Association found that one out of five Ivy League college students was taking "smart drugs" to help improve academic performance. By 2015, that number had jumped to one in three (in all college students). Almost immediately, there was a backlash. You might think the backlash was about safety. After all, the term *"smart drug"* applies to the unsupervised and often dangerous off-label use of ADHD drugs like Ritalin and Adderall. But public health wasn't the issue.

Instead, in November 2015, *USA Today, Washington Post,* and a half-dozen other major news outlets all asked the same question: "Are Smart Drugs Cheating?" It's a peculiar question. Students use these drugs because they improve focus and help them work longer and harder. It's the same thing a cup of coffee and a study group provides. But gatherings of caffeinated coeds aren't considered cheating, so what makes smart drugs any different?

Or consider our distrust of a more controversial class of "smart" drugs: psychedelics. Seventy years ago, the influential University of Chicago historian Mircea Eliade coined the phrase "archaic techniques of ecstasy" to describe singing, dancing, chanting, and meditation, or all the "original and pure" techniques that shamans used to alter consciousness. But he left out an important category. While shamans on nearly every continent have long utilized psychotropic plants like mushrooms and cacti to shift states and access insight, Eliade omitted this fact, editing history to come down on the side of the skin-bag. "Narcotics," he argued in his classic book *Shamanism,* "are only a vulgar substitute for 'pure' trances . . . an imitation of a state that the shaman is no longer capable of attaining otherwise."

So whether we're talking about students popping pills or shamans taking psychedelics, the bias is the same. It's a question of effort. Studying all night for an exam takes work. Adderall feels like cheating. The same goes for grueling hours of drumming, chanting, and meditation versus the near-certain transformation produced by mind-altering plants.

The Pale of the Body is ascetic to its core: no pain, no gain. Altered states that arise within ourselves, via internal catalysts like prayer and meditation, are considered stable, reliable, and earned. If the goal is genuine transformation, then nothing as fleeting or pleasant as a flow state or psychedelic session can substitute for decades of prayer and meditation. "The ultimate wisdom of enlightenment," author Sam Harris emphasized in his recent bestseller *Waking Up,* "whatever it is, cannot be a matter of having fleeting experiences. . . . Peak experiences are fine, but real freedom must be coincident with normal waking life."

In other words, insights gleaned from within the skin-bag are valid and true, while those glimpsed outside the skin-bag are not to be trusted. Experiences that require external catalysts, such as psychedelics and smart drugs, are volatile, unreliable, and, ultimately, too easy.

In 1962, attempting to settle this skin-bag debate, Walter Pahnke conducted one of the more famous psychedelic experiments in history. A graduate student in theology at Harvard Divinity School, Pahnke gathered a group of twenty seminary students at Boston University's Marsh Chapel on Good Friday. To see if mind-altering drugs could produce "authentic" mystical experiences, he gave half the group psilocybin, the other half the active placebo niacin (which produces similar physiological changes without the cognitive effects), then everyone went into chapel to attend the Good Friday service.

Afterward, subjects rated the service for a variety of mystical qualities: sacredness, ineffability, distortion of time and space, and a sense of oneness with the divine. "[Psilocybin] subjects ranked their experiences much higher in mystical qualities than members of the control group did," explains John Horgan in his book *Rational Mysticism.* "Six months later, the psilocybin group reported persistent beneficial effects on their attitude and behavior; the experience had deepened their religious faith. . . . The experiment was widely hailed as proof that psychedelic drugs can generate life-enhancing mystical experiences." So life-enhancing, in fact, that nine out of the ten seminary students who received psilocybin ended up becoming ministers, while none of the placebo group stayed on the path to ordination.

Yet the skin-bag bias has been hard to shake. Despite coming from a tightly controlled study at one of the nation's top institutions, these findings didn't do much to alter popular opinion or academic consensus. Researchers have twice gone back to validate

Pahnke's work. In 2002, Johns Hopkins psychopharmacologist Roland Griffiths got the same results when he reran the experiment with full double-blind modern standards. When author Michael Pollan asked him about this unusual need for redundancy, in a 2015 *New Yorker* article, Griffiths's answer said it all: "There is such a sense of authority that comes out of the primary mystical experience that it can be threatening to existing hierarchical structures. We end up demonizing these compounds. Can you think of another area of science regarded as so dangerous and taboo that all research gets shut down for decades? It's unprecedented in modern science."

Yet it's important to remember that the skin-bag bias isn't simply about our distrust of pharmacology—whether study drugs or psychedelics. It's really about our distrust of technology in general. That was cyborg philosopher Andy Clark's point. And, since consciousness-altering technology is changing fast, these developments are bringing up new test cases for what we consider "legitimately earned" ecstasis.

Consider Laurentian University neuroscientist Michael Persinger's God Helmet. More than fifty years ago, researchers discovered that electrical stimulation of the right temporal lobe can produce visions of God, sensed presences, and other notable altered states. Persinger built a helmet that directs electromagnetic pulses toward this brain region. More than two thousand people have tried out the device, and the majority have had some form of nonordinary experience.

Already, commercial versions of the God Helmet are available online, as are stories of DIY hackers who are reproducing its basic effects with little more than some wires and a nine-volt battery. There's talk about developing a version for virtual reality and incorporating it in video games.

Other researchers are pushing neurotech even further. Palo Alto Neuroscience, a Silicon Valley start-up, has developed a system that can tag the biomarkers of a nonordinary state—that is, brainwaves, heart rate variability, and galvanic skin response—and then use neurofeedback to guide you back there later. Trained meditators like Tibetan monks can put themselves into a transcendental state, and the machine will record their profile. Soon, as the technology matures, a novice will be able to put on the device and use these biomarkers to steer toward the same experience.

But if we continue to insist that smart drugs and psychedelics are cheating, what happens as the boundaries between ourselves and our tools continue to blur? As technological upgrades and modifications to our inner state become increasingly common, what happens to the Pale of the Body when whole swatches of the populace begin finding God in the machine?

The Pale of the State

In 2008, a middle-aged woman walked into David Nutt's office in Bristol, England. Nutt is a psychiatrist and psychopharmacologist, specializing in the treatment of brain trauma, and this woman was in need of help. A serious head injury had caused a dramatic shift in her personality. She'd completely lost the ability to feel pleasure, becoming impulsive, anxious, and occasionally violent. Things had gotten so bad that she could no longer work, her children had been placed into foster care, and even her local pub had had enough—banning her for life after she'd begun abusing the staff.

Nutt was no stranger to serious head trauma, but most of the cases he saw involved drug abuse. This woman didn't use drugs. Her injury had been sustained while horseback riding. Like most people, Nutt had assumed that horseback riding was a safe outdoorsy pastime. Yet when he checked the data, he was surprised to discover the sport produced serious injury or death in one out of every 350 outings.

At the time, Nutt was also the chair of the British Advisory

Council on the Misuse of Drugs. Part of his job involved assessing and ranking the harms of various substances and reporting those findings to the government and the public. As this was the late 2000s, the substance then getting the most attention was MDMA. Fueled by the rave scene, the drug had spread across England like wildfire. The press had been talking about it in epidemic terms. Politicians were vilifying it as public enemy number one. Nutt wasn't so sure.

After meeting that woman, he'd done a back-of-the-envelope calculation comparing the injuries and deaths caused by horseback riding—which he dubbed "equasy"—to those produced by MDMA. But even when he discounted the downstream costs of drug use such as addiction, violent behavior, and traffic accidents, his numbers showed the dangers of *equasy* and *ecstasy* several orders of magnitude apart. For every 60 million tablets of MDMA consumed, Nutt found 10,000 adverse events, or one for every 6,000 pills popped. He then compared that number to the 1-in-350 tally for horseback riding and published the results.

Headlines across the country read: "British policy doctor claims ecstasy is safer than riding a horse." Tabloids had a field day. The internet picked up the story and soon both houses of Parliament were hotly debating the topic. Within a week, Nutt was called before the home secretary (somewhere between the U.S. attorney general and the head of homeland security) and was publicly chastised for his irresponsible and incendiary remarks.

Yet, for Nutt, what he was saying wasn't incendiary, it was simply the facts. "Ecstasy is a harmful drug," he explained in his 2012 bestseller, *Drugs—Without the Hot Air*. "But how harmful? As harmful as drinking five pints of beer? As harmful as riding a motorbike? David Spiegelhalter, a professor of risk communications [at Cambridge University], has calculated that taking an ecstasy pill is as dangerous as riding a motorbike for about six miles or a [pedal] bike for twenty miles. These sorts of comparisons are useful because they can help people make choices about their behavior based on realistic assessments of the risks. Politicians, however, are highly resistant to them."

And that's putting it mildly. Consider this Abbott and Costello–like "Who's on first" exchange between Nutt and the home secretary:

Home Secretary: *You can't compare harms from an illegal activity with a legal one.*

Nutt: *Why not?*

Home Secretary: *Because one's illegal.*

Nutt: *Why is it illegal?*

Home Secretary: *Because it's harmful.*

Nutt: *Don't we need to compare harms to determine if it should be illegal?*

Home Secretary: *You can't compare harms from an illegal activity with a legal one.*

After the hearing, it was clear to Nutt that the British government wasn't interested in data-driven comparisons. But that calculation had got him thinking. He decided to assess twenty of the most common substances of abuse for nine different categories of harm, including physical, mental, and social impact.

A quick scan of Nutt's evidence-based rankings confirm what many would suspect. On the list of toxic substances, drugs like heroin, crack, and methamphetamine rank high. No question about it, they're really bad for you and really bad for those around you. But, while heroin is so destructive it claims the number-two slot, it still couldn't beat out the number-one scourge: alcohol. And tobacco—another legal staple of modern life—clocked in at number six, two ahead of marijuana, and just behind cocaine and methamphetamine. And what about MDMA, that supposed public enemy number one? It barely made the list, coming in at number 17, just ahead of LSD and magic mushrooms, which were 18 and 20 respectively. So, while those substances are arguably our most "feared" drugs, when Nutt examined the facts, they weren't even close to the most "harmful."

Nutt told the public about this work in a lecture at King's College London. This time, the combination of the sensational subject matter and the media's endless appetite for top-ten lists created the perfect viral storm. Everyone from the *Guardian* to *The Economist* picked up the story. The press crunched the data down to a headline: "Government minister claims alcohol more dangerous than LSD!"

Nutt was back on the hot seat. This time it was too hot. A Home Office spokesman told the public: "The Home Secretary expressed surprise and disappointment over Professor Nutt's comments which damage efforts to give the public a clear message about the dangers of drugs." A few days later, Nutt was relieved of his drug czar post, forever becoming "the scientist who got sacked."

But this does raise an important question: why did Nutt lose his job? After all, he'd been hired to provide an evidence-based message about the harmfulness of drugs and had done just that. In fact, he'd done such solid work that *The Lancet*, one of the most reputable medical journals in the world, published his findings. But the issue had nothing to do with the quality of his research. Nutt had transgressed a different barrier, the Pale of the State.

In very simple terms, the states of consciousness we prefer are those that reinforce established cultural values. We enshrine these states socially, economically, and legally. That is, we have *state-sanctioned states of consciousness*. Altered states that subvert these values are persecuted, while the people who enjoy them are marginalized.

Take Ritalin and Adderall, the ADHD meds that students as young as grade school pop like candy. These drugs don't even make an appearance on Nutt's list, while methamphetamines claim a top five spot. But they're essentially the same substance.

"Aside from some foul cutting material," explains journalist Alexander Zaitchik in *Vice*, "Winnebago methamphetamine and pharmaceutical amphetamine are kissing cousins. The difference between them boils down to one methyl-group that lets crank race a little faster across the blood-brain barrier and kick just a little harder. After that, meth breaks down fast into good old dextroamphetamine, the dominant salt in America's ADHD and cram-study aid, Adderall."

Yet, our attitudes toward these substances—their inside or outside the pale status—is markedly different. The 1.2 million Americans who tried meth last year were breaking bad, while the 4.4 million American *children* who took ADHD drugs were striving to become better students. Same drugs, different contexts. One is manufactured by major pharmaceutical companies and enthusiastically dispensed by suburban doctors; the other is cooked up in trailers and sold on street corners.

Or consider three substances that sit squarely inside the state's pale: caffeine, nicotine, and alcohol. The coffee break, smoke break, and happy hour are the most culturally enshrined drug rituals of the modern era, even though two of the three are top-ten offenders in Nutt's rankings. There's hardly a single workplace in the Western world that doesn't, at least informally, support this triad. And for good reason. An optimally tuned market economy needs alert employees who work as hard as possible for as long as possible. So dedicated time-outs for stimulant consumption (that is, the coffee break and, these days, the e-cigarette break) are institutionally sanctioned and socially reinforced.

Which is where the cocktails come in. Without the soothing effects of alcohol, the cigarettes-and-coffee workforce would become jittery wrecks within a fortnight. Add in some booze from time to time and you've got a finely tuned cycle of stimulation-focus-decompression that dovetails with broader economic goals. "In the competitive environment of the firm," explains Intel researcher and author Melissa Gregg in the *Atlantic*, "it is little wonder that workers resort to performance-enhancing drugs. . . . When so many jobs require social networking to maintain employability, these mood enhancers are a natural complement to the work day after 5 p.m. In an always-on world, professional credibility involves a judicious mix of just the right amount of uppers and downers to remain charming." Because these substances drive us forward, they continue to sit inside society's perimeter fence, and never mind the evidence.

And that fence is the real reason Nutt lost his job. Even though the information he presented was considered, medical, and factual, it went against established norms and policies. It threatened approved channels of awareness and the substances that support them. Nutt ventured beyond the Pale of the State and ended up, professionally, burned at the stake.

Pipers, Cults, and Commies.

Hamelin is a town of about fifty thousand people, nestled among the rolling hills of central Germany. The buildings are timber-framed sandstone, the lanes are narrow and winding, the beer gardens are cozy. And then there are the rats. In Hamelin, the rats are everywhere.

In photos, on paintings, depicted in the stained glass of the town's eight hundred-year-old church. All the bakeries sell them: rat-shaped cakes, cupcakes, and loaves of bread. They're available as keepsakes and key chains. They are the town's most famous citizens. Their tale dates back a thousand years, retold by Goethe, the Brother's Grimm, and the poet Robert Browning, a warning to parents and children alike.

According to the "Luneburg Manuscript," the only written account of the actual event, in 1284 Hamelin was suffering a rodent infestation. That's when a wandering minstrel with a magic flute showed up. He claimed to be a rat catcher, willing to rid the town of its problem, but for a fee. The locals agreed to his price and the piper went to work. He played his flute and entranced the rats. They followed him wherever he went: out of the town's main gates, down to the river, and into the water, where, unable to break free from the power of the music, they drowned.

It was a job well done; the issue was the fee. The locals refused to pay the piper. So he stormed off, vowing revenge. A few months later he returned, but this time, when he played his flute, it wasn't the rats who followed—it was the children.

The citizens of Hamelin recorded their loss in their town register, afterward dating all proclamations according to the years and the days since the tragedy. Even today, the Hamelin town hall still bears the inscription:

> In the year 1284 after the birth of Christ
> From Hamelin were led away
> One hundred and thirty children, born at this place.
> Led away by a piper into a mountain.

Historians continue to debate the tale of the Pied Piper of Hamelin. Early interpretations held that the rats were a carrier of plague, and this was an outbreak story. Others have argued it was a tale of forced conscription and a Children's Crusade. A few scholars have focused on a singular detail—that magic flute whose tune none can resist—and argued this might actually be a story about the irresistible attraction of music, dance, and trance, against which a stern medieval church could not compete. So while we usually tell the Pied Piper story as a morality play, a reminder to pay your debts and keep your word, it might actually be a warning about the lure of ecstasis.

This is no idle warning. History is littered with tales of ecstatic explorations gone wrong. Consider the 1960s. Ken Kesey snuck LSD out of a Stanford research lab and all manner of tie-dyed hell broke loose. The same thing happened with the sexual revolution of the 1970s. What began as a quest for personal liberation ended up in spiking rates of marital dissatisfaction and divorce. And 1990's rave culture too, which blended synthetic drugs with electronic music, collapsed under a series of tightening legal restrictions, ER visits, and tabloid fodder.

Which brings us to the final reason the Stealing Fire revolution has remained hidden from view: nearly every time we light out into this terrain, somebody gets lost. By definition, ecstasis makes for tricky navigation. The term means out of our heads and "out" isn't always pleasant. These states can be destabilizing. It's why psychologists use terms like *"ego death"* to describe the experiences. "[It's] a sense of total annihilation," writes psychiatrist Stanislav Grof in his book *The Adventure of Self-Discovery.* "This experience of ego-death seems to entail an instant merciless destruction of all previous reference points in the life of an individual." In short, Alice didn't wander into Wonderland—she *fell* down the rabbit hole.

Making matters worse, these experiences are enticing. Sometimes we revisit them more often than we should. The $4 trillion of the Altered States Economy is a stark testament to the depth of that desire. So while we've painted the guardians of the pale in a somewhat reactionary light, let's give the gatekeepers their due. What lies beyond the pale isn't always safe and secure. Outside the fence of state-sanctioned consciousness, there are, to be sure, peaks of profound insight and inspiration. But there are also the swamps of addiction, superstition, and groupthink, where the unprepared can get stuck.

For this reason, most people don't venture outside alone. We look for others who have gone this way before us; we look for guidance and for leadership. But, as the Pied Piper's story illustrates, not everyone who leads us beyond that fence has our best intentions at heart.

This past century was thick with cautionary tales. Bhagwan Shree Rajneesh and his bioterrorist followers, Marshall Applewhite and the Heaven's Gate suicides, and Charles Manson and the Tate-LaBianca murders are well-known examples. There are plenty of others. Combine enticing experiences with clay-footed gurus and you have a recipe for disaster.

No wonder parents of the 1960s hugged their children close as they traipsed off to California (or Bali or Maui) with flowers in their hair. There really was no telling if the next enlightened sage was a huckster, a demagogue, or both. Better to never venture out the door than roll those dice. Isn't that why Harvard professor Timothy Leary, whose greatest crime was telling undergrads to "tune in, turn on, and drop out," ended up branded by President Richard Nixon "the most dangerous man in America"?

And it's not just the unscrupulousness of leaders; it's the power of the tools they wield. During ecstasis, our sense of being an individual "I" gets replaced by the feeling of being a collective "we." And this doesn't just happen in small groups like the SEALs on night ops or Googlers at a desert festival. It's also the feeling that arises at large political rallies, rock concerts, and sporting events. It's one of the reasons people go on spiritual pilgrimages, and why evangelical megachurches are booming (with more than six million attendees every Sunday). Bring a large group of people together, deploy a suite of mind-melding technologies, and suddenly everyone's consciousness is doing the wave.

"*Communitas*" is the term University of Chicago anthropologist Victor Turner used to describe this ecstatic sense of unity. This feeling tightens social bonds and ignites enduring passion—the kind that lets us come together to plan, organize, and tackle great challenges. But it's a double-edged sword. When we lose ourselves and merge with the group, we are in danger of losing too much of ourselves. Our cherished rational individualism risks being overrun by the power of irrational collectivism. This is how the ideals of the French Revolution veered into the bloody mob rule of the Reign of Terror. It's why, Turner argued, communitas is too potent to unleash without proper checks and balances: "Exaggerations of communitas, in certain religious or political movements of the leveling type, may be speedily followed by despotism."

In the 1930s, Adolf Hitler provided a frightening example, co-opting traditional techniques of ecstasy—light, sound, chanting, movement—for his Nuremberg rallies. "I am beginning to comprehend some of the reasons for Hitler's astounding success," wrote Hearst journalist William Shirer in 1934. "Borrowing a chapter from the Roman church, he's restoring pageantry . . . and mysticism to the drab lives of twentieth century Germans." Hitler wasn't just borrowing from Rome, but from the United States as well. According to Fuhrer confidant Ernst Hanfstaengl, "the 'Sieg Heil' used in political rallies was a direct copy of the technique used by American college football cheerleaders. American college type music was used to excite the German masses who had been used to . . . dry-as-dust political lectures."

Hitler wasn't the only twentieth-century despot to rely on these techniques. Stalin, Mao, and Pol Pot sold the same thing: a Utopia of We, the experience of communitas at scale. They even sold it the same way. Nearly identical stump speeches: "Individualism is out. We are all one. No one is better than anyone else. Anyone who disagrees will either be shot, imprisoned or 'rehabilitated.'" As Nietzsche said: "madness is rare in individuals—but in groups, political parties, nations and eras, it's the rule." And in ecstatic groups, it's practically unavoidable.

So why have we missed a revolution in human possibility?

Because altered states have a distinctly checkered history of altering (nation) states. Because pipers, cults, and commies scare the pants off us. Because the drive to get out of our heads has ended in tragedy as often as ecstasy. Because the pale protects us as much as it confines us. Because no one wants to end up like the children of Hamelin, lured beyond the safety of the town walls, never to be heard from again.

PART TWO

The Four Forces of Ecstasis

No one dances sober, unless he is insane.

—*Cicero*

CHAPTER FOUR

PSYCHOLOGY

Translating Transformation

In the Middle Ages, priests routinely complained that their congregants nodded off in church. Still, despite the stuffy pews and unintelligible sermons, parishioners made sure they were awake for the most interesting part of the service: the miraculous transformation of bread and wine into the body and blood of Christ. At that exact moment, the priest would utter a powerful incantation, what sounded like "hocus pocus," and the magic would be done. Except, as Archbishop of Canterbury John Tillotson later noted, "in all probability . . . *hocus pocus* is nothing else but a corruption of *hoc est corpus* ("this is the body"), [a] ridiculous imitation of the priests of the Church."

Without the ability to understand Latin, the peasants had garbled the translation, providing would-be magicians with a catchphrase for centuries to come. But in 1439, Johannes Gutenberg invented the printing press, giving the faithful the chance to read the Bible in their own language. Rather than deferring to the clergy, laypeople could suddenly debate and interpret scriptures and draw their own conclusions. This broke the church's hold on Renaissance Europe and paved the way for reformations and revolutions.

Something similar is happening today. Thanks to accelerating developments in four fields—psychology, neurobiology, pharmacology, and technology; call them the "Four Forces of Ecstasis"—we're getting greater access to and understanding of nonordinary states of consciousness. These forces give us the chance to study, debate, reject, and revise long-standing beliefs of our own. We're becoming more precise in our translations, learning to rely less on hocus-pocus and superstition, and more on science and experience.

Advances in psychology have given us a better sense of our own development and, with it, the space to move beyond socially defined identity. Stepping outside the monkey suits of our waking selves no longer means risking ridicule or madness. Higher stages of personal development have been demystified. We now have the data-driven models needed to navigate this formerly obscure terrain and clearer frameworks to make sense of the journey.

Advances in neurobiology, meanwhile, have clarified our understanding of what is happening in our brains and bodies when we experience a range of mental states. This sharpened perspective allows us to strip out the interpretations of past gatekeepers and understand, in simple and rational terms, the mechanics of transcendence. And unlike the take-it-on faith dictates of traditional mythologies, the discoveries of neurobiology are testable.

Pharmacology gives us another tool to explore this terrain. By treating the six powerful neurochemicals that underpin ecstasis as raw ingredients, we've begun to refine the recipes for peak experience. We're translating the cookbook for *kykeon,* allowing us to tune these states with increasing precision, and giving us access to them on demand.

Our last force, technology, brings that access to scale. Whether we're relying on flow-producing neurofeedback or awe-inducing virtual reality, these breakthroughs turn once-solitary epiphanies into experiences that can be shared by hundreds of thousands of people at once. More people having more experiences means more data and firmer conclusions.

Combined, these forces give us unprecedented insight into the upper range of human experience. In Part Two of this book, we're going to examine them in detail, seeing where they come from and why they matter. We'll meet the innovators and experts at the forefront of this movement, an unlikely collection of digital artists, consciousness hackers, sex therapists, and molecular chemists—to name a few—who are harnessing these four forces to drive change in the real world. By democratizing access to some of the more controversial and misunderstood territory in history, these modern-day Gutenbergs are taking experiences once reserved for mystics and making them available to the masses.

The Bell Tolles for Thee

In February 2009, Oprah Winfrey teamed up with Eckhart Tolle for *Oprah and Eckhart: A New Earth,* a ten-part online video series devoted to Tolle's nontraditional ideas about spirituality. Eleven million people from 139 different countries tuned in to watch. Global brands—Chevy, Skype, and Post-it—sponsored the series. *A New Earth* attracted 10 million people, or 800,000 more than turned up in New York City for the pope's last visit and 9 million more than the largest haj (pilgrimage to Mecca) on record—making this webcast one of the ten largest spiritual "gatherings" in recorded history.

Yet Tolle remains an unusual candidate for guru of the century. He grew up in the rubble of postwar Germany, suffering acute anxiety and severe depression. By age ten, he was contemplating suicide. By age twenty-nine, while a doctoral candidate in philosophy at the University of Cambridge, he snapped.

Tolle dropped out of graduate school and spent the next two

years homeless, mostly sitting on a park bench in central London. He passed the time in a state of near-constant bliss—a state of oneness with the universe—that, he maintains, persists to this day.

In spiritual terms, Eckhart Tolle found sudden enlightenment. In the language of this book, he stabilized ecstasis, making the temporary selfless, timeless, and effortless experience of a non-ordinary state a part of his permanent reality. But go back a decade or two and traditional psychiatrists might have assessed his case very differently. Tolle would have been kept in a padded cell, sedated with Thorazine, and given a steady dose of electroshock therapy. Instead, he teamed up with Oprah to beam his unassuming talk of nondual consciousness out to millions of seekers around the world.

What Tolle is preaching is nothing less than the Gospel of STER. His core argument is that through the experience of selflessness, timelessness, and effortlessness—his so-called "Power of Now"—we can dwell in a place of unlimited richness. And, if the popularity of his webcast is anything to go by, this idea is resonating with millions of people.

Which brings us to an important question: How did any of this happen? How did we get from a man who would have been institutionalized as clinically insane a few decades ago to where we are today, with that same man leading one of the largest spiritual gatherings in history?

To answer this question, we'll need to understand how psychology became a force for ecstasis. We'll start with origin stories, seeing how the human potential movement broadened the vocabulary of inner experience and then brought that vocabulary to the mainstream. A glimpse under the covers of a twenty-first-century sexual revolution will explore how an expanding menu of acceptable practices has given more people access to non-ordinary states than ever. We'll then examine how researchers are using peak experiences to cure trauma in terminal patients and survivors of abuse and war. Finally, we'll see how scientists have begun to integrate ideas about ecstasis into a rigorous model of human psychology that shows altered states don't just make us feel better for a moment, they can actually further our development over a lifetime. But, before we can do any of that, we have to turn back the calendar some seventy years, and meet ourselves as we once were.

Mad Men

Coming out of World War II, our concept of self—of who we actually were—didn't stretch very far. With the cardboard cutouts of Organization Man on one side and Betty Homemaker on the other, our sense of ourselves was constricted almost to the point of caricature. Hollywood stars like Gary Cooper and John Wayne epitomized a "strong and silent" masculine ideal, while soap operas and advertisements sold a flawless stay-at-home femininity. This was the era of *The Man in the Gray Flannel Suit,* when suburban conformity, consumerism, and corporate ladder-climbing had come to signify success.

But all this started to change in the late 1950s, as the Beats' passionate rebellion found its voice. "The Beat Generation was a vision that we had," explained Jack Kerouac in *Aftermath*, "of a generation of crazy illuminated hipsters suddenly rising and roaming America . . . characters of a special spirituality . . . staring out the dead wall window of our civilization." Allen Ginsberg's epic poem *Howl* was a shout out that same window: a free-verse rant about the need to break loose from social constraint via direct, primal experience.

Nowhere did this urge for raw self-expression show up more visibly than at Esalen, the Big Sur, California–based institute that the *New York Times* once called the "Harvard of the Human Potential movement." So central was Esalen to the evolving identity of that generation that the popular TV show *Mad Men* ended with its main character Don Draper experiencing an epiphany on its oceanside front lawn (and as a harbinger of the spiritual marketplace that would soon emerge, he promptly turned that insight into the iconic "I'd like to buy the world a Coke" ad campaign).

While histories of Esalen tend to focus on founder Michael Murphy, whose family had owned that idyllic stretch of California coastline for generations, the path that led to Eckhart on *Oprah* was largely laid by Dick Price, Esalen's cofounder and first director.

In 1952, Price came out of Stanford with a degree in psychology, and went to Harvard for graduate school, but, frustrated by the conservatism of the faculty, dropped out. He moved to San Francisco, bumped into the Beats, and, under their tutelage, began exploring Eastern mysticism and primal experience. Shaken loose from his moorings, Price suffered a manic episode at a North Beach, San Francisco, bar in 1956 and was hauled off for a three-month stint on an Army psych ward.

While the state labeled him "psychotic," Price didn't accept that he had gone crazy. He labeled his unhinging a "transitional psychosis" and argued that his time on the ward served a useful purpose: unlocking a door within himself. "[My] so-called 'psychosis,'" Price later said, "was an attempt toward spontaneous healing, and it was a movement toward health, not a movement toward disease."

Price's realization—the idea that we sometimes have to "break down to breakthrough"—quickly became a mainstay of the human potential movement. It's one of the reasons we can now view Tolles's park bench madness as a spiritual initiation rather than as a psychological meltdown.

Over the next two decades, Price and Murphy evolved this insight into a pragmatic philosophy. They took the best that organized religion had to offer, stripped out anything that was doctrinal or impractical, and placed a heavy emphasis on ecstatic experimentation. It was a "pragmatic culture of sensation and know-how," notes author and modern religious historian Erik

Davis in *AfterBurn*, "an essentially empirical approach to matters of the spirit that made tools more important than beliefs. Consciousness-altering techniques like meditation, biofeedback, yoga, ritual, isolation tanks, tantric sex, breathwork, martial arts, group dynamics and drugs were privileged over the claustrophobic structures of authority and belief that were seen to define conventional religion."

It was a uniquely American approach that resonated deeply with the country's anti-authoritarian ideals. Rice University religious scholar Jeff Kripal called it the "Religion of No Religion," writing in his book *Esalen,* "It has no official alliance with any religious system. It can provide, like a kind of American Mystical Constitution, a spiritual space where almost any religious form can flourish, provided . . . that it does not . . . claim to speak for everyone. As an early Esalen motto put it, 'No one captures the flag.'"

Despite its inclusive charter, the institute's impact was limited to a thin slice of the population. With its deep roots in Ivy League intellectualism and ascetic Eastern mysticism, this new philosophy mainly attracted well-educated freethinkers from New England and California. That is, until Werner Erhard came along and transplanted Esalen's esoteric ideas out of bohemia and into the belly of the mainstream.

Erhard, a self-educated car salesman from St. Louis, was drawn to personal development through his study of motivation. He quickly realized that many of the ideas of the human potential movement had applications beyond spiritual seeking. So Erhard repackaged an assortment of Esalen-inspired practices into a business-friendly format, creating EST, short for the Erhard Seminars Training. The seminar deliberately reproduced Price's accidental transformation, engineering a "breakdown-to-breakthrough" experience via a series of marathon, fourteen-hour days, without food or breaks, and with lots of yelling and profanity—the fabled "EST encounter."

Prior to Erhard, most spiritual seekers skewed anti-establishment and antimaterialist—which is fine if you're living in a monastery or off a trust fund, but problematic if you need to make a living. And even more problematic if you're trying to sell seminars. In the same way that Henry Ford realized his workers had to be able to afford one of his Model T's for his company to thrive, Erhard understood that seekers needed to be financially successful enough to afford his next workshop. So he hitched the human potential movement to the wagon of the Protestant work ethic. Napoleon Hill's *Think and Grow Rich* replaced the *Bhagavad Gita* as seminal text. Mandalas were out. Vision boards were in. And the American spiritual marketplace has never been the same.

If you've ever hired a personal or executive coach (professions that didn't exist before the late 1970's), heard someone say they "just needed space," been encouraged to "take a stand" or "make a difference," or engaged in a journey of "transformation" around your "personal story"—you've come across terms coined or popularized by Erhard and his trainings. And with this expansion of vocabulary and emphasis on material success, what had been formerly "other," that is, couched in the esoteric language of Asian religions and incompatible with the American dream, became us—part of our everyday vernacular, and accessible to all.

While EST itself made an impact, with almost one million people going through those original seminars, Landmark, the latest incarnation of Erhard's teachings, boasts corporate clients including Microsoft, NASA, Reebok, and Lululemon. Personal development, which only a few decades ago had been mocked and marginalized, has become a credible way to "optimize human capital" at some of the country's most successful organizations.

We hear echoes of these ideas everywhere, from Tony Robbins's empowerment seminars to the prosperity theology preached every Sunday by megachurch ministers like Joel Osteen. And even though Tolle's teachings stay remarkably grounded in his own realization, the larger notion that spiritual awareness can bring material fulfillment—that, perhaps, you can have nirvana, a happy marriage, *and* the shiny new car—helps explain how 11 million fans of Oprah even heard of him in the first place.

Taking the Kink Out of Kinky

The lineage that goes from Esalen to EST to Eckhart is one of increasing self-exploration, of pushing beyond the limits of what was considered safe or acceptable. Price and the Beats gave us a way to get past the taboos of primal expression and mental illness. Erhard broke down the taboo segregating spiritual merit from material success. The human potential movement normalized the use of ecstatic practices for psychological growth. Along the way, we came up with broader versions of ourselves, and new ways to interact with each other. But nowhere were taboos more visibly challenged than in the realm of sexuality.

And while the sexual revolution of the 1960's and '70's increased the *amount* of sex people were having, we want to focus here on a more recent development—what *kind* of sex people are having today. Specifically, how a growing subset of experiences formerly defined as "kinky"—meaning taboo and outside the norm—are giving more people access to ecstasis than ever.

In 2014, we were invited to San Francisco to speak about the overlap between the neuroscience of flow, meditation, and sexuality, and see the cutting edge of this latter domain up close. Justine Dawson, the CEO of OneTaste and our host for the weekend, escorted us to our chairs in the front row of a packed auditorium, ascended the stage, dropped her pants, and lay back on a massage table.

OneTaste's founder, Nicole Daedone, entered stage right. Wearing a gray wool dress and a large black apron, she snapped on a pair of latex gloves, dipped her forefinger and thumb into a jar of artisanal lube, and went to work. The reclined Dawson began mewling. With an entertainer's flair for the theatrical, Nicole paused, swiveled on a stilettoed black boot, and punched her hand in the air like a rock guitarist. The audience began calling out words to describe their own experience. "Tingling in my groin," one woman announced. "Heat," said another. "Tumescence," blurted a software engineer.

On the OneTaste website, they describe their central practice as OMing, short for "orgasmic meditation," and we'd just witnessed a mainstage demonstration by the masters. A tightly circumscribed, almost ritualized practice, OMing involves stroking the upper left quadrant of a woman's clitoris for exactly fifteen minutes without attachment to outcome or expectation of reciprocity. Their goal is to create a "turned on" woman—one who is neurochemically saturated, physically open, and emotionally empowered.

And they're not the first, by any means, to use sex as a trigger for nonordinary states of awareness. From the ancient "wine, women, and song" to the boomers' "sex, drugs, and rock and roll," erotic techniques have always featured heavily on Promethean playlists.

"The search for personal transformation, including through sex, led to the oceanside hot tubs at . . . Esalen," explains Patricia Brown in her *New York Times* article on orgasmic meditation. "One Taste is but the latest stop on this sexual underground, weaving together strands of radical individual freedom, Eastern spirituality, and feminism."

Their message appears to be getting traction: OneTaste has centers in Los Angeles, New York, San Francisco, London, and Sydney, along with a dozen other cities. They've received largely favorable coverage in the *Atlantic, New Yorker, Vanity Fair, Time,* and dozens of other publications (the clincher is always a female journalist who tries it out herself). To put this in perspective, Planned Parenthood founder Margaret Sanger had to flee the United States in 1914 to avoid prosecution for sharing basic information about contraception. Yet, in 2015, OneTaste notched a placement on the Inc. 5000—an annual ranking of high-growth companies.

The popularity of orgasmic meditation makes sense once you understand what it can do. “In French literature,” University of Pennsylvania neurologist Anjan Chatterjee explains in his book *The Aesthetic Brain,* “the release from orgasm is famously referred to as *la petite mort*, the little death . . . the person is in a state without fear and without thought of themselves or their future plans. . . . This pattern of deactivation could be the brain state of a purely transcendent experience enveloping a core experience of pleasure.”

Social scientist Jenny Wade has spent her career studying these same phenomena. “The fact is, sex—all by itself,” she writes in her book *Transcendent Sex,* “can trigger states identical to those attained by spiritual adepts of all traditions.” By Wade’s estimate, nearly 20 million Americans have had at least one encounter with boundary-dissolving, self-obliterating sex. “[It’s] happened to countless thousands of people regardless of their background,” she notes, “to hairdressers, investment managers, nurses, lawyers, retailers and executives.”

But if 20 million of us have experienced transcendent sex, why aren’t we talking about it more often? “Most volunteers said they had never confided their experiences to their lovers,” Wade reports, “for fear their partners would make fun of, not be interested in, or not be receptive to ‘spiritual stuff.’” Yet it’s not only the “spiritual stuff” that gives people pause. Sometimes it’s merely who’s doing what with whom, and what would the neighbors think.

For most of the twentieth century, we had no idea what people were up to in their bedrooms or what normal actually looked like. Long after Masters and Johnson and Kinsey and company did their best to extract honest answers from modest people about their sex lives (and were consistently lied to—men overestimated the length of appendages, women sandbagged on their number of partners, and everyone fudged when reporting edgier behavior), we still tend to keep our most intimate thoughts and experiences under wraps. But technology is helping to lift that veil, creating a "digital commons" where we can discreetly bypass nosy neighbors to explore our actual desires.

For the past five years, the French research group Sexualitics has been building the "Porngram," an analytic tool to track online sexual behaviors across the world. "Traces left by billions of users give us cultural snapshots of tastes," they wrote in their 2014 paper "Deep Tags: Toward a Quantitative Analysis of Online Pornography." "[M]ore importantly, they enable researchers to look for structures and patterns in the evolutionary dynamics of practices adopted by a significant and growing proportion of the human population."

And the biggest pattern revealed by this data is that we're more curious about the outer edges of human sexuality than at any other time in history (and this includes women, who now make up a third of all online porn viewers). Simply judging by today's top search terms—an assortment that would make a sailor blush—we find a greater interest in activities that, until recently, were considered deviant by psychiatric professionals or prosecuted outright by law.

Take, for example, BDSM, short for bondage, discipline, sadism, and masochism, a category that includes a range of intense pleasure/pain stimulation and role-play. Until recently, BDSM was mostly practiced by a fringe subculture, and at some risk. In America, evidence of these behaviors was admissible in court as grounds for divorce or denial of child custody.

But, in 2010, the American Psychiatric Association agreed to redefine "aberrant sexual behavior" in the *Diagnostic and Statistical Manual of Mental Disorders* (the bible for mental illness classification), making a clear distinction between elective play and actual pathology. For the first time, consenting adults weren't considered morally deviant or mentally ill for choosing sexual behavior that was beyond the pale.

Which was just in time, considering the following year E. L. James published *Fifty Shades of Grey*. A critically panned but wildly popular novel, *Fifty Shades* cataloged the BDSM-driven sexual awakening of a college student at the hands of a handsome young billionaire. It became one of the fastest-selling books in history and sold more copies on Amazon than all of J. K. Rowling's seven-volume Harry Potter series combined. But, if it wasn't literary merit that vaulted this book into a global phenomenon, what was it?

You could make a case that, in the same way that electronic dance music thrives largely because of its ability to generate a shift in state, *Fifty Shades* represents a sexual equivalent. It's a de facto user manual for erotic ecstasis that many of its readers never dared imagine. With this one book, Kindles and minivans replaced brown-paper-wrapped smut and trench coats. Edgy sex went suburban.

If you get past the snickering, the exceptional popularity of the book is understandable. Beyond simple novelty or variety, this broader range of sexual experiences is giving people access to altered states that "vanilla" sexuality rarely does. To return to the "knobs and levers" of ecstasis, we know full-spectrum sexuality contains many of the same triggers that produce STER through meditation, flow states, and psychedelic experiences. Pleasure produces endorphins, but pain can prompt even more. The uncertainty of teasing, as Stanford's Robert Sapolsky established, spikes dopamine 400 percent. Nipple stimulation boosts oxytocin. Pressure in the throat or colon regulates the vagus nerve, creating exhilaration, intense relaxation, and goose bumps, what Princeton gastroenterologist Anish Sheth memorably terms *poo-phoria*. "To some it may feel like a religious experience," Sheth writes, "to others like an orgasm, and to a lucky few like both."

And momentary erotic transcendence can bring lasting change. A 2013 Dutch study found that kinky sex practitioners "were less neurotic, more extraverted, more open to new experiences, more conscientious, less rejection sensitive, and had higher subjective well-being." Nor are these benefits reserved just for the socially progressive. Minister Ed Young of the mega–Fellowship Church in Dallas, Texas, exhorts his congregation of thousands to the "Seven Days of Sex" challenge, where they commit to sex every day for a week to deepen their spiritual union. "And when I say intimacy, I don't mean holding hands in the park or a back rub," Young clarified. "If you make the time to have sex, it will bring you closer to your spouse and to God."

This grassroots movement, combined with growing research, make it clear that sexuality is becoming one of the more popular techniques of ecstasy available today. Once we step beyond taboos and social conditioning, it's easier to see that access to ecstasis has been hardwired into us all along. In the past, you had to risk social or legal censure if you tried to figure it for yourself. Now there's a broad movement to explore full-spectrum sexuality and elevate it from compulsion or perversion, into something more deliberate, playful, and potent.

The arc of the moral universe may be long, but it's bending toward the kinky.

Good for What Ails You

Even with all these new approaches to ecstasis at our disposal—from the EST encounter to more adventuresome sexuality—a critical question remains: can these fleeting moments produce meaningful psychological change? If not, it would be hard to justify their additional risks and complications. If so, they might earn a legitimate spot in the tool kit of personal development. And one of the starkest tests for ecstasis is, can it help survivors of serious trauma? If you can heal them, presumably, you can heal just about anyone.

One of the first people to try to address this question was Brown University neuroscientist Willoughby Britton. In the 1990s, Britton became interested in near-death experiences (NDEs), where subjects had transcendental encounters during life-threatening events (including the widely reported tunnel-of-light journey). Thirty years of research showed that people who had an NDE scored exceptionally high on tests of overall life satisfaction. As a trauma expert, Britton found this unusual.

In her experience, most people who came close to dying were scarred by the event, developing post-traumatic stress and other mental health conditions. But if those studies were right, then people who had a transcendent near-death experience were having a decidedly atypical response to trauma.

Britton decided to find out how atypical. After recruiting twenty-three NDE'ers and twenty-three control subjects, she hooked them all up to electroencephalogram (EEG) machines and ran a sleep study. Her goal was both to get a clear picture of brainwave activity and record how long it took her subjects to enter REM sleep—an excellent way to measure happiness and well-being.

Normal people go into REM at about 90 minutes; depressed people enter sooner, usually at 60 minutes. Generally happy people head in the opposite direction, dropping into REM at around 100 minutes. Britton discovered that NDEers delayed entry until 110 minutes—which meant that they were off the charts for happiness and life satisfaction.

And when Britton examined the EEG data, she discovered why: the brain-firing patterns of her subjects were completely altered. It was as if the NDE had instantly rewired their gray matter. Sure, it was only a single experiment, but it did suggest that even a one-time encounter with a powerful altered state could impart lasting change.

Despite this intriguing result, few studies followed. NDE's posed a serious research challenge: they are, *ipso facto*, accidental, and not easily repeatable in the lab.

Then Johns Hopkins neuroscientist Roland Griffiths came up

with an elegant solution. Rather than scouring the country for small groups of NDE survivors, Griffiths tapped a much larger population facing death: terminal cancer patients. And instead of waiting for a fleeting and hard to replicate tunnel-of-light journey, he relied on chemistry to produce a similar impact on-demand.

In 2011, Griffiths gave three grams of psilocybin to a group of terminal cancer patients, in an attempt to provide them with relief from fear-of-death anxiety (which is understandably hard to alleviate). Afterward, he administered a battery of psychological tests, including a standard fear-of-dying metric, the Death Transcendence Scale, at one- and fourteen-month intervals. Just as with Britton's NDE survivors, Griffiths found significant, sustained change: a marked decrease in their fear of death, and a significant uptick in their attitudes, moods, and behavior. Ninety-four percent of his subjects said taking psilocybin was one of the five most meaningful experiences of their lives. Four out of ten said it was *the* most meaningful.

More recently, researchers have learned that you don't have to be on the verge of dying to experience relief—that an encounter with ecstasis can help the "walking wounded" as well. In 2012, psychologist Michael Mithoefer discovered that even a single dose of MDMA can reduce or cure post-traumatic stress disorder (PTSD) in survivors of child abuse, sexual abuse, and combat. "It was completely cathartic," reported an Army ranger who suffered a broken back and severe head trauma in Iraq. "The next day [after just one session] the nightmares were gone. I was glowing and extroverted for the first time since getting blown up. MDMA gave me my life back."

To put the options trauma survivors have into perspective, nearly 25 million Americans suffer from PTSD, yet the only two drugs approved for treatment are Prozac and Zoloft. Both require weeks or even months to get into our system, while their effects last only as long as we continue taking them. Stop the pills and you return, more or less, to where you started.

In contrast, Mithoefer found that the benefits provided by one to three rounds of MDMA therapy lasts for years. These results outstrip conventional treatments so convincingly that, in May 2015, the federal government approved studies of MDMA as a treatment for depression and anxiety.

Flow researchers have achieved comparable results without drugs, simply by altering neurobiological function. In 2007, working with Iraq War veterans at Camp Pendleton, occupational therapist Carly Rogers of the University of California, Los Angeles blended surfing (a reliable flow trigger) and talk therapy into a treatment for PTSD. It was essentially the same protocol Mithoefer used, only with the flow generated by action sports substituting for MDMA.

Much like in Mithoefer's study, sufferers experienced nearly immediate relief. "After just a few waves, they [the soldiers with PTSD] were laughing in the surf lineup," reported *Outside* magazine. "'Oh my God, our Marines are talking,' said the lieutenant who approved the experiment. 'They don't talk. Ever.'"

Since then, the program has been formalized, and more than a thousand soldiers have taken part. Hundreds of veterans and surfers have volunteered their time, including eleven-time world champion Kelly Slater. And their investment paid off. In a 2014 paper published in the *Journal of Occupational Therapy,* Rogers reported that after as little as five weeks in the waves, soldiers had a "clinically meaningful improvement in PTSD symptom severity and in depressive symptoms."

And surfing isn't the only non-pharmacological intervention to show promise. A recent study done by the military found that 84 percent of PTSD subjects who meditated for a month could reduce or even stop taking selective serotonin reuptake inhibitors (SSRIs). In contrast, the control group—who didn't meditate and stayed on antidepressants—experienced a 20 percent *worsening* in PTSD symptoms during that same period.

Taken together, all this work—from the NDE studies to

the cancer and trauma research to the flow and meditation programs—demonstrates that even brief moments spent outside ourselves produce positive impact, regardless of the mechanisms used to get there. And they can provide these benefits in the face of the deepest challenges imaginable.

Yet, in each of these instances, the only people given permission to explore altered states were those, quite literally in some cases, left for dead. It's almost as if dispensing these techniques to normal people would be unseemly or, at the very least, unscientific.

In her book *Train Your Mind, Change Your Brain,* science writer Sharon Begley highlights this problem, describing the history of psychology as one favoring remediation over transformation: "Science has always focused . . . on people and conditions that are pathological, disturbed, or at best normal. . . . In the past thirty years, there have been about forty-six thousand scientific studies on depression and an underwhelming four hundred on joy. . . . As long as someone can obtain nonsickness that is deemed sufficient. As Buddhist scholar Alan Watts put it, 'Western scientists have an underlying assumption that normal is absolutely as good as it gets and that the exceptional is only for saints, that it is something that cannot be cultivated.'"

But many of the same interventions that can help us get our heads above water can just as effectively be devoted to raising our heads above the clouds. If we're interested in untapped levels of performance improvement and lasting emotional change, peak states of consciousness may provide the quickest path between two points: a shortcut from A to E(cstasis).

Altered States to Altered Traits

One summer day, while working in the garden with his young daughter, Nicki, University of Pennsylvania psychologist Martin Seligman had, in his own words, "an epiphany." Seligman was meticulously freeing weeds with a trowel, and neatly setting them aside in a discard pile. Nicki, being five, was just having fun. "Weeds were flying up in the air," Seligman later said, "dirt was spraying everywhere."

Seligman, who describes himself as both a "serious gardener" and a "serious grouch," couldn't take it. He started yelling. But Nicki wasn't having any of it. She stomped over with a stern look on her face.

"Daddy," she said, "I want to talk with you. From the time I was three until I was five, I whined a lot. But I decided the day I turned five to stop whining. And I haven't whined once since. . . . If I could stop whining, you can stop being such a grouch." Seligman decided to take her up on the challenge, and bring the field of psychology along for the ride.

In 1998, after being elected president of the American Psychological Association, Seligman made positive psychology the central focus of his tenure. "I want to remind our field that it has been side-tracked," he wrote in his first Presidential Column for the APA's newsletter. "Psychology is not just the study of weakness and damage, it is also the study of strength and virtue. Treatment is not just fixing what is broken, it is nurturing what is best within ourselves."

If those trauma studies demonstrated that a few instances of ecstasis can help mend what's broken, then what happens if we deploy these techniques repeatedly, over the course of a lifetime? Can recurring access to these states really "nurture what is best within ourselves?" Can they, as Alan Watts suggested, be used to "cultivate the exceptional"?

Oddly, in the history of adult psychology, the idea that we could cultivate anything over time was considered suspect. After adolescence, the thinking went, adults were pretty much fully baked. Sure, we could learn technical skills, like going to business school or picking up a musical instrument, but our ability to add psychological capacities—like the gratitude and empathy that Nicki asked her father to embrace—was believed to be pretty much over and done with by the time we'd graduated from college.

But Bob Kegan, the Harvard psychologist whom we first met in Chapter 2, upended that assumption by doing something psychologists before him hadn't done too much of: longitudinal research. Kegan tracked a group of adults as they aged. His goal was simple: understand how they changed and grew over time, and determine if, in fact, there were upper limits to who we can become.

Kegan spent three decades tracking this group, seeing what happened to their psychological maturity and capacity along the way. He discovered that while some adults remained frozen in time, a select few achieved meaningful growth. Right around middle age, for example, Kegan noticed that some people moved beyond generally well-adjusted adulthood, or what he called "Self-Authoring," into a different stage entirely: "Self-Transforming."

Defined by heightened empathy, an expanded capacity to hold differing and even conflicting perspectives, and a general flexibility in how you think of yourself, self-transforming is the developmental stage we tend to associate with wisdom (and Roger Martin's *Opposable Mind*). But not everyone gets to be wise. While it usually takes three to five years for adults to move through a given stage of development, Kegan found that the further you go up that pyramid, the fewer people make it to the next stage. The move from self-authoring to self-transforming for example? Fewer than 5 percent of us ever make that jump.

But in all of this developmental research, buried in the footnotes about those self-transcending 5 percenters, lay a curious fact. A disproportionate number of them had dabbled in ecstasis: often beginning with psychedelics and, after that, making meditation, martial arts, and other state-shifting practices a central part of their lives. Many of them described their frequent access to non-ordinary states as the "turbo-button" for their development.

And this isn't an isolated finding. Fifty years ago, psychologist Abraham Maslow noticed that the more peak experiences a person had, the closer they came to self-actualization, his term for the upper stages of adult development. A 2012 study published in *Cognitive Processing* took it further. When examining the relationship between peak experiences and performance in Olympic athletes and corporate managers, researchers found that the highest performers didn't just have more frequent peak experiences; they also made more ethical and empathetic decisions.

Boston College's Bill Torbert found that those at the top of the developmental pyramid not only were more ethical and empathetic; they performed better in the workplace as well. In a survey of nearly five hundred managers in different industries, he found that 80 percent of those who scored in the upper two stages of development held senior management roles despite only making up 10 percent of the broader population. The most developed leaders, as Torbert noted in the *Harvard Business Review*, "succeeded in generating one or more organizational transformations over a four-year period, [and] their companies' profitability, market share, and reputation all improved." Consciousness, it turns out, goes straight to the bottom line.

If the shift in psychology that led us from Esalen to Eckhart was about greater permission to explore, then Kegan and his colleagues have given us the next piece of that puzzle: a map of where we're going. By bridging the gap between peak states and personal growth, these discoveries validate ecstasis as a tool not only for self-discovery, but also for self-development. So while ecstatic states (which are brief and transitory) aren't the same as developmental stages (which are stable and long-lasting), it appears that having more of the former can, under the right conditions, help accelerate the latter. In short, altered *states* can lead to altered *traits*.

CHAPTER FIVE

NEUROBIOLOGY

Outside the Jar

There's a bit of southern folk wisdom that says "you can't read the label while you're sitting inside the jar." And that notion, that we can't always understand what we're too close to, sums up the relationship between psychology and neurobiology as forces for ecstasis. As substantial as the advances in psychology have been, what they've really done is make the inside of the jar bigger—by expanding our sense of what's possible. But the field of neurobiology is doing something else altogether. By giving us an understanding of the ingredients on the label, it's providing a view of our lives from outside the jar.

In the past, we might have seen all of our psychological ups and downs as challenges to be solved with our minds. Now we can address them at a more foundational level. With a clearer view of the knobs and levers being tweaked in the body and brain, neurobiology provides us with a more precise tool kit with which to tackle life's challenges.

To trace these developments, we'll take a look at how Hollywood movie stars clued us into the link between motion and emotion, how webcams and video game sensors are helping revolutionize mental health, how the U.S. military and top business schools are harnessing biometrics to predict the future, and how maverick scientists are reverse-engineering ancient mystical experiences. Taken together, these examples will show how neurobiology has given us the tools to map and measure what's happening in our bodies and brains when we're experiencing both the ordinary and the extraordinary. And the results are changing how we think about how we think.

I Can't Feel My Face

In February 2011, Nicole Kidman wowed on the Academy Awards' red carpet with a silver Dior gown, a 150-carat Fred Leighton diamond choker, and matching Pierre Hardy pumps. Yet, beyond her upscale designer choices, Kidman drew attention that evening for an unexpected reason. Her eyebrows. She couldn't *not* arch them. She looked like a porcelain doll who'd just sat on a pinecone.

And while Kidman repeatedly ascribed her timeless beauty to diet, exercise, and sunscreen, in 2013 she finally admitted to Italy's *La Repubblica,* "I did try Botox, unfortunately, but I got out of it, and now I can finally move my face again." Which is kind of critical if you're an actor who earns a living making faces.

Not long after celebrities began showing up at gala events with implausibly blank expressions, researchers started to notice that Botox was doing more than just altering how people looked. It was altering how they felt, too. In study after study, when seriously depressed patients received Botox injections in their frown lines, they got significant and sometimes instantaneous relief from depression. But when Botoxed subjects were asked to empathize with other people, to feel their joy or share their sorrow, they simply couldn't.

This struck scientists as strange. Since the time of the Greeks, Western thinkers have considered the mind the engine that drives the bus and the body the passenger that comes along for the ride. It's the mind-body split, a one-way arrow of causation that insists the head is always in charge (and can be trusted to govern our higher aspirations), while the body is the vessel that houses our animal instincts (and should be strictly controlled). But these Botox studies pointed in the opposite direction. Somehow, changes in the body—freezing the face with a neurotoxin—were producing changes in the mind: the ability to feel sadness or empathy. The horse appeared to be steering the rider.

And we now know why. Our facial expressions are hardwired into our emotions: we can't have one without the other. Botox lessens depression because it prevents us from making sad faces. But it also dampens our connection to those around us because we feel empathy by mimicking each other's facial expressions. With Botox, mimicry becomes impossible, so we feel almost nothing at all. No wonder Nicole Kidman was relieved to get a few wrinkles back.

But the bigger point is that these studies reflect a sea change in how we think about thinking. They move us from "disembodied cognition," the idea that our thinking happens only in the three pounds of gray matter tucked between our ears, to "embodied cognition," where we see thinking for what it really is: an integrated, whole-system experience. "The body, the gut, the senses, the immune system, the lymphatic system," explained embodied cognition expert and University of Winchester emeritus professor Guy Claxton to *New York* magazine, "are so instantaneously and complicatedly interacting that you can't draw a line across the neck and say 'above this line it's smart and below the line it's menial.'"

In fact, we're not smart and we have bodies—we're smart *because* we have bodies. The heart has about 40,000 neurons that play a central role in shaping emotion, perception, and decision making. The stomach and intestines complete this network, containing more than 500 million nerve cells, 100 million neurons, 30 different neurotransmitters, and 90 percent of the body's supply of serotonin (one of the major neurochemicals responsible for mood and well-being). This "second brain," as scientists have dubbed it, lends some empirical support to the persistent notion of gut instinct.

And these whole-body perceptions can be easily influenced. If someone gives you a cup of icy cold water to hold, then introduces you to a stranger, as researchers at Yale did, you'll treat this newcomer with suspicion and rate them as colder and more distant on personality scales. But if they give you a cup of hot coffee and make the same introduction, trust comes more easily. The act of feeling physical warmth is enough to trigger a cognitive change: you literally warm up to people, no thinking required.

Or consider Harvard psychologist Amy Cuddy's popular TED talk about the power of body language. Cuddy discovered that spending two minutes in a "power pose"—meaning a posture of dominance (like "Wonder Woman": hands on hips, elbows cocked wide, legs firmly planted)—changed both psychology and physiology. In her research, subjects who adopted the Wonder Woman posture took greater risks and took them more frequently. And two minutes of the pose was enough to increase levels of the dominance hormone testosterone by 20 percent and decrease the stress hormone cortisol by 15 percent. While the field of embodied cognition is in its infancy, and there is still lots of work to be done replicating studies and integrating insights, these early findings suggest a tighter linkage between our minds and our bodies than most of us would ever suspect.

So what does any of this have to do with ecstasis? For those interested in shifting states, knowing that the body can drive the mind gives us a whole new set of knobs and levers with which to play. Einstein's quote "you cannot solve a problem at the level at which it was created" is invariably used to encourage higher, more expansive solutions. But the opposite is equally true. Sometimes, lower, more basic solutions can have just as big an impact.

In our work training flow, for example, we've found that action and adventure sport athletes deliberately amplify the physical sensations of gravity to help shift their mental state. Whether it's boosting g-forces by carving hard turns on skis or mountain bikes, or nullifying g-forces with jumps, spins, and airs, these athletes expand the range of normal bodily inputs to push themselves into the zone. "Weightlessness, weightedness and rotation are the nectar of gravity games," explains professional climber and filmmaker Jimmy Chin. "They provide easy access to flow, and that's what keeps us coming back for more."

Really, none of this is new. Five thousand years ago, early yoga practitioners were tinkering with embodied cognition to prompt higher states of awareness. If simply standing like Wonder Woman for a few minutes is enough to produce meaningful changes in our hormonal profile, imagine what practicing a full sequence of yoga postures every morning would do. "There's all this evidence that [movement sequences] have an impact on stress," Peter Strick, a professor at the University of Pittsburgh's Brain Institute, writes in *Proceedings of the National Academy of Sciences.* "it has an effect on how you project yourself and how you feel."

Fifteen hundred years ago, Shaolin monks in China became the ultimate Buddhist warriors by training their bodies to elevate their minds. They spent years practicing nearly impossible physical feats—smashing bricks with their hands, stopping spears with their necks, balancing all their weight on two fingers—as a way of training their minds. In direct contrast to skin-and-bone ascetics who sought ecstasis by ignoring or denying the body, these monks believed transcendence began with its total mastery.

In the West, though, we inherited a different legacy, that mind-body split. It began with a healthy dose of Judeo-Christian guilt—that our bodies were not to be trusted—and was cemented by an increasingly industrial economy, where our bodies were less and less needed. And today, with so much of our emotional and social lives mediated by screens, we've become little more than heads on sticks, the most disembodied generation of humans that has ever lived.

But if we draw upon the insights of embodied cognition research, we can reconnect our bodies and brains. We can shift posture, breathing, facial expressions, flexibility, and balance as a way to tune our states of consciousness, altered or otherwise. We don't have to process everything first and foremost through our psychology. We can flip the script entirely and change our experience without having to think much at all. Funk master George Clinton once sang, "free your mind and your ass will follow," but he might have had it backward. Free your ass and your *mind* will follow.

AI Shrink

In 2014, we traveled to the University of Southern California to meet an entirely new kind of therapist who's taking the insights of embodied cognition and turning them inside out. Rather than using bodily changes to impact mood, she's measuring bodily expression to unearth deeper psychological conditions.

Her name is Ellie. She's a professional woman in her early thirties, with olive skin, brown eyes, and brown hair worn in a ponytail. She dresses demurely, in a blue scoop-necked shirt, a brown cardigan sweater, and a teardrop pearl around her neck. In conversation, her questions are straightforward and inquisitive. "Where are you from?" "How are you feeling today?" Occasionally, as when Steven tells her he's from Chicago, Ellie reveals a bit of personal information. "Oh," she says, "I've never been there. I'm from L.A., myself."

And while that much is true—Ellie is, technically, from Los Angeles—she was less born there, than built there. She's the world's first artificially intelligent (AI) shrink. Created by researchers at DARPA and USC's Institute for Creative Technologies, Ellie's designed to identify signs of depression, anxiety, and

PTSD in soldiers. She's part of a larger Defense Department initiative to identify mental health concerns earlier and stem the tide of suicide in the military.

Aside from the fact that Ellie appears only on a video monitor, a session with her is about what you'd expect from a traditional therapist. She begins each appointment with rapport-building questions, such as "How are you feeling today?" She asks direct follow-ups—"When was the last time you felt really happy?"—and weaves in clinical inquiries: "How much sleep are you getting?" Sessions close with queries intended to boost patients' mood: "What are you most proud of?"

Beneath the surface, though, Ellie is anything but a traditional therapist. Her ability to identify, assess, and respond to emotion in real time is the result of a growing body of research into the mechanics of embodied cognition. The neurobiology of emotion shows that our nonverbal cues—our tics, twitches, and tone—reveal much more about our inner experience than words typically do. "People are in a constant state of impression management," explains USC psychologist Albert "Skip" Rizzo, the director of the institute. "They have their true self and the self they want to project to the world. And we know the body displays things that sometimes people try to keep contained."

While the research that led to Ellie required advanced brain imaging and a DARPA budget, her ability to track a patient's unconscious tells involves inexpensive, off-the-shelf technology: a Logitech webcam to monitor facial expressions, a Microsoft Kinect movement sensor to follow gesture, and a microphone to capture word choice, modulation, and inflection. Every second, she's noting and processing more than sixty different data points. She constantly scans vocal tone for signals of sadness, for example, with every word rated on a seven-point "openness" scale (that is, willingness to disclose revealing information). An array of algorithms then analyzes this data and helps provide a clearer picture of a patient's overall well-being.

"Ellie's the third leg of the stool," explains Rizzo. "For the past century, scientists only had good data about two of the three streams of information we can glean from people. There's what people say about themselves, self-reporting, and what the body can tell us, biophysical data like heart rate and galvanic skin response. But there's also behavior—our movements and facial expressions. These have always been hard to assess and, typically, we could only get at them through subjective observations. Ellie gathers objective, high-quality data."

With her cameras, sensors, and algorithms, she extends our five senses and gets upstream of our umwelts—or reality as we perceive it. She bypasses our relentless storytelling and reflects back to us a little more of what we're actually thinking and feeling.

And people prefer talking to Ellie than to actual humans. Even trained psychologists tend to judge. Ellie never does. In a 2014 study, the USC team discovered that patients were twice as likely to disclose personal information to her than to a human therapist. The researchers' ultimate goal is to make her available via laptop, to anyone with a Wi-Fi connection.

AI therapists like Ellie are simultaneously more objective and more perceptive than humans and they can help us become the same. She gives us distance from our inner critic and a better understanding of what we're perceiving in the present moment. In a very real sense, Ellie's dispassionate reflection of who we are mimics the advantages conferred by ecstasis—the ability to look at ourselves from outside ourselves.

So, when we next find ourselves thrust into the selflessness of a nonordinary state, ripped free of our waking identity and its comfortable narratives, it won't be as shocking or disorienting. Ellie's dispassionate reading of our biometrics gives us more accurate insight than our own self-reporting. She can help us close the gap between what Rizzo calls our "true selves" and "the self [we] want to project to the world."

Precognition Is Here (But You Knew That Already)

In 1999, Steven Spielberg had a problem: how to translate Phillip K. Dick's short story "Minority Report" into a box-office hit. A cop thriller set in the mid-twenty-first century, "Minority Report" hinges on mutant humans who can see into the future, using their "precognition" to prevent crimes before they happen. Unfortunately for Spielberg, Dick's book contained few clues as to what this Tomorrowland world actually looked liked. So the director pulled together a collection of the world's best futurists to help him color in the storyboards in the most believable way.

Jaron Lanier, author and virtual reality pioneer; Shaun Jones, the first director of DARPA's Unconventional Countermeasures program; and the heads of both the Center for Bits and Atoms and the Media Lab at the Massachusetts Institute of Technology, all convened at Shutters, a Santa Monica, California, beachside hotel, for their secret think tank. While few moviegoers remember the intricacies of *Minority Report*'s plot, nearly everyone remembers the world they created—mostly because they got so much of the future right. Out of all the ideas Spielberg's think tank came up with, they correctly anticipated head-up digital displays, gesture command, driverless cars, personalized advertising, voice-automated home appliances, and predictive crime analytics.

At first, the movie was supposed to be set in 2070, but then they dialed it back to 2054. And out of all their educated guesswork, that was their biggest mistake. They were forty years too late. And not just for the cool tech like driverless cars and head-up displays, but for precognition itself.

In 2015, we were invited to Southern California to host an event for Nike's innovation group. Every year, to help spark new insights in their own design work, this group goes on creative field trips to learn more about the cutting edges of other fields. That year, Nike wanted to learn about precognition and group flow and how harnessing the former could get their teams more of the latter. So we designed a think tank of our own. In conjunction with colleagues from Advanced Brain Monitoring, as well as some DARPA researchers and MIT Media Lab hackers, we set out to give Nike a glimpse of a future that was already here, but not yet widely distributed.

If you think back to the embodied cognition work of Amy Cuddy, AI Ellie, and others, their big insight was that our bodies, facial expressions, posture, and voice all convey more information than we would ever suspect. And, if we change any of those things, we can substantially shift how we feel and what we think in the present moment. That's pretty big news. But what we explored with Nike went even further than that, beyond "real-time" transformation and into "future-time" prediction—precognition itself.

Chris Berka, the founder of Advanced Brain Monitoring, kicked off the day, briefing the group on a few of her team's research projects, starting with one they'd recently completed with the U.S. Navy. The project involved submariners, often considered the roughest job in the military. Getting locked into a tin can for six months at a time, never seeing sunlight, and carrying warheads that could start World War III requires a special kind of sailor. It also requires tightly coordinated teamwork and a lot of group flow. The trouble was, it had been next to impossible for the Navy to predict who could weather those conditions and still be able to drop into the zone, and who might end up stabbing someone with a fork at dinner.

So that was the puzzle Berka's team set out to solve. First, the Navy built a full-scale dryland replica of a nuclear-class submarine. Next, Advanced Brain Monitoring took teams of submariners, hooked them up to EEG machines to track brainwaves and heart rate variability (HRV) monitors to measure the quality of their cardiac rhythms, and then put them through a sixteen-week training program that simulated all the stresses of actual deployment.

Simply by examining the biometric profile of each sailor, Berka and her team could reliably track and identify which sailors were going to excel at collaborative problem solving. With nothing more than these data printouts, they could tell the difference between a novice team still fumbling around, a sophomore team just starting to gel, and a high-performing team of experts. By the end of the full sixteen-week program, they could predict—months in advance of actual deployment—which teams were going to be able to "flip the switch" and drop into group flow together, and which ones ran the risk of flipping out a thousand feet underwater. By measuring biometrics in the present, they were literally able to see into the murky depths of the future.

And the armed forces aren't the only ones interested in finding better ways to predict what's around the bend. In a related study run in Barcelona, Spain, researchers at the ESADE Business School (twice ranked the top business school in the world by the *Wall Street Journal*) wanted to know if they could identify "emergent leadership" long before their students graduated. So Berka and her team took thirty-five MBA students, hooked them up to EEG and HRV monitors, and gave them a case study to solve.

Once again, by examining the neurophysiological profile of these students, Berka was able to identify "emergent leaders"—those individuals who would have an outsize positive impact on the team and its decision making—in as little as thirty minutes. While there was no correlation between emergent leadership and how much students talked or even what they said, there was a direct relationship between their neurophysiological responses and those of their classmates. Transformational leaders not only regulated their own nervous systems better than most; they also regulated other people's.

In the same way that multiple clocks on a wall end up synchronizing to the one with the biggest pendulum, emergent leaders can entrain their entire teams and create a powerful group flow experience. In this study at ESADE, this shared experience helped the groups arrive at more creative and ethical solutions (as rated by a panel of faculty and experts). Furthermore, these young leaders' ability to create group coherence proved to be a reliable indicator of effective decision making later in their careers. Until now, this was a latent, invisible capacity, not predictable, and definitely not trainable.

Now, thanks to advances in neurobiology, we can map and develop these ineffable skills with little more than some basic sensors and algorithms. Our understanding of the science has progressed to the point where we can not only shift how we think and feel in the present, but also make accurate predictions about how we're going to think and feel in a future that has yet to occur.

The Birth of Neurotheology

Dr. Andrew Newberg doesn't look like a rebel. With his curly brown hair, boyish face, and easy manner, he's more soccer dad than academic cultural outlaw. Nothing about his demeanor suggests that this was the guy to overturn a hundred years of scientific thinking. But that's exactly what happened.

Back in the early 1990's, science and religion weren't on speaking terms. Serious researchers didn't study spirituality. Newberg felt differently. A neuroscientist at the University of Pennsylvania, he had a deep interest in the fundamental nature of reality and the differences between the world that we perceive and the world as it really is. And this led him to reconsider mystical experiences—especially the experience known as "unity."

"Unity," Newberg explains, "is the feeling of being one with everything. It's a foundational notion in pretty much every religious tradition. There are thousands of depictions of the experience. And if you read through them, you'll find that people often describe unity as more 'fundamentally real' than anything else they've ever experienced. More real than reality. Well, what does that mean? I think it means that in trying to answer this question we need to take into account both the science and the spirituality, that we can't just dismiss the latter because it makes us uncomfortable as scientists."

Newberg's inquiry coincided with rapid advancements in neural imaging technologies like functional magnetic resonance imaging (fMRI) and positron emission tomography (PET). With these developments, researchers went from trying to understand the structure of the brain to trying to understand its function. Questions like what is the brain doing when we laugh or shop or have an orgasm were suddenly answerable. Newberg thought that if those experiences could be decoded, why not mystical experiences? Why not unity?

At the time, these were controversial questions. "When I got interested in the subject," continues Newberg, "very few scientists thought mystical experiences were real. Telling a shrink you felt one with everything was a good way to get locked in a padded cell and trying to do research on it was an easy way to get denied tenure."

But Newberg persisted, becoming the first person to use advanced brain imaging technology to examine mystical experiences. For research subjects, he chose Franciscan nuns and Tibetan Buddhists. During moments of intense prayer, the nuns report *unio mystica,* or oneness with God's love. Peak meditation, meanwhile, brings the Buddhists into "absolute unitary being" or, as they say, "oneness with the universe." By using single-photon emission computed tomography (SPECT) to take pictures of their brains during these sublime moments, Newberg tested those claims.

The scans showed significant deactivation in the right parietal lobe, a key component in the brain's navigation system. This part of the brain helps us move through space by judging angles and distances. But, to make these judgments, this region must first decide where our own body ends and the rest of the world begins, essentially drawing a boundary line between self and other.

It's an important boundary. People who suffer a stroke or

brain damage to this area struggle to sit down on a couch because they don't know where their leg ends and the sofa begins. It's also a flexible boundary. When race car drivers feel the road beneath their pedals, or blind people feel the sidewalk through the tips of their cane—or, for that matter, when SEALs merge with their team on a night op—it is partially the result of the right parietal lobe blurring the boundary of self.

What Newberg discovered is that extreme concentration can cause the right parietal lobe to shut down. "It's an efficiency exchange," he explains. "During ecstatic prayer or meditation, energy normally used for drawing the boundary of self gets reallocated for attention. When this happens, we can no longer distinguish self from other. At that moment, as far as the brain can tell, you are one with everything."

In finding biology beneath spirituality, Newberg helped bridge the gap between science and religion. For the first time, mystical experiences were understood not as a symptom of mental illness or divine intervention, but rather as the by-product of normal brain function. Almost overnight, an area once off-limits to researchers opened for exploration. It was the birth of the field known as neurotheology—the application of the tools of modern brain science to the study of religious experience.

And unity is only the first in a long series of those experiences that researchers have now decoded. "It's amazing how far neurotheology has come," explains Newberg. "Different types of meditation, chanting, singing, flow, prayer, mediumship, speaking in tongues, hypnosis, trances, possession, out-of-body-experiences, near-death experiences, and sensed presences—they've all been examined using high-powered imaging."

A good way to get a sense of neurotheology's rapid progress is to examine the work of Dr. Shahar Arzy, director of the neuropsychiatry lab at Hebrew University in Jerusalem and one of the researchers following in Newberg's footsteps. While Newberg was intrigued by one of the more widely reported mystical experiences (unity), Arzy, in 2011, became interested in one of the rarest: the doppelganger, or seeing a vision of one's double.

Back in the thirteenth century, the Jewish mystic Abraham Abulafia, the founding father of Ecstatic Kabbalah, developed a meditation system supposedly able to produce this experience. Combining an elaborate series of instructions, this method not only created a doppelganger, but, Abulafia claimed, allowed you to ask it questions and seek guidance.

Building off Newberg's work on the parietal lobe, Arzy theorized that a region right next door—the temporal-parietal junction—might be responsible for this doppelganger effect. By integrating data about vision, touch, balance, and time, this part of the brain answers the question, Where am I right now? Abulafia's method, Arzy suspected, was specifically designed to scramble this function.

The Kabbalist's formula combines breathing, visualization and prayer and with a series of head movements: when you pronounce the first letter of God's name, slowly tilt your face to the sky; when you pronounce the second letter, thrust your neck backward and head forward like a woodpecker. "The temporal-parietal junction sits right atop the region that processes motion," explains Arzy. "It's possible these head movements trick this region into sending out body position error messages. This could compound the [state-changing] effects of the visualization and meditation and produce a doppelganger."

Arzy confirmed his hunch when he was able to use fMRI to examine an epileptic girl who had been seeing visions of her double. The scan revealed damage to her temporal-parietal junction, providing a neurological explanation for a formerly mystical experience. He then found a way to precipitate this phenomenon in healthy people. By using virtual reality to force viewers to toggle perspective back and forth between two versions of themselves, he created a high-tech update to Abulafia's meditation. And it worked. Almost instantaneously, study subjects could no longer tell their real selves from their simulated doppelgangers—no belief or practice required.

In the fifteen years since Newberg founded the field of neurotheology, we've gone from initial investigations of these effects to being able to reproduce them at will. Both developments provide a clearer understanding of the mechanics of ecstasis—essentially a Clif Notes version of esoteric techniques that took thousands of years to evolve. Consider that most religions have lengthy recipes for encountering the divine. Hundreds of ingredients: what to eat, what to wear, whom to marry, how to act, what to believe, and, of course, what kind of spiritual practices to perform. But out of that entire list, there's only a fraction of "active ingredients" that reliably impact brain function and alter consciousness.

Neurotheology lets us validate which ingredients actually make a difference. "In unexpected ways," writes David Brooks in the *New York Times,* "science and mysticism are joining hands and reinforcing each other. That's bound to lead to new movements that emphasize self-transcendence but put little stock in divine law or revelation. Orthodox believers are going to have to defend particular doctrines and particular biblical teachings. . . . We're in the middle of a scientific revolution. It's going to have big cultural effects."

Possibly very big cultural effects. Because this work is nondenominational, it speaks to a wider swath of people than established orthodoxies ever could. Certainly, atheists have used the fact that there's neuronal function beneath mystical experience to claim that spirituality is merely a trick of the brain. But neurotheology takes a faith-neutral position. All this work proves is that these experiences are biologically mediated. If you're a believer, it offers a deeper understanding of divine methods. If you're a nonbeliever, it provides another consciousness-altering tool upon which to draw. Either way, these advances do more than just provide an academic explanation for the ecstatic—they provide a user manual on how to get there.

OS to UI

Abraham Maslow once famously said, "When all you've got is a hammer, every problem looks like a nail." What he meant was, when it comes to problem-solving, we tend to get locked into using familiar tools in expected ways. The technical term for this is the *Law of the Instrument.* Give someone a hammer and, indeed, they'll look for nails to pound. But present them with a problem where they need to repurpose that same hammer as a doorstop, or a pendulum weight, or a tomahawk, and you'll typically get blank stares.

We may be facing a similar situation when it comes to our minds. At least as far back as the French Enlightenment and Descartes's *cogito ergo sum* (I think, therefore, I am), we've relied on our rational selves—what psychologists call our "egos"—to run the whole show. It's a Maslow's hammer kind of reaction. Every issue we encounter, we try to solve by thinking.

And we know it's not working. Even a quick glance at today's dire mental health statistics—the one in four Americans now on psychiatric medicines; the escalating rate of suicide for everyone from ages ten to seventy-eight—shows how critically overtaxed our mental processing is these days. We may have come to the end of our psychological tether. It might be time to rethink all that thinking.

With the recent advancements in neurobiology, we now have options: Embodied cognition teaches us that how we move our bodies affects our brains and minds. AI therapy proves that our subconscious expressions can reflect our inner state more accurately than we do. Precognition demonstrates that we can anticipate how we're going to feel and think in the future by tracking (and even altering) our biometrics in the present. Neurotheology integrates all of these findings and lets us reverse-engineer a whole host of nonordinary states, just by working backward from our neurophysiology.

Rather than treating our psychology like the unquestioned operating system (or OS) of our entire lives, we can repurpose it to function more like a user interface (or UI)—that easy-to-use dashboard that sits atop all the other, more complex programs. By treating the mind like a dashboard, by treating different states of consciousness like apps to be judiciously deployed, we can bypass a lot of psychological storytelling and get results faster and, often, with less frustration.

Take, for example, one of the most common ailments of the modern world—mild to moderate depression. Instead of moping around, hoping for things to get better on their own, we can scan our UI and choose an alternate program to run. We could get on a treadmill (studies show exercise is effective for depression in all but severe cases), or get some natural sunshine (70 percent of Americans are deficient in vitamin D, which has a direct impact on mood), or practice meditation for fifteen minutes (a paper in the *Journal of the American Medical Association* found it as effective as SSRI's and without the side effects). None of these approaches require thinking about our thinking, but each of them can significantly shift our mood.

Choices like these are available not just in our personal lives, but in our professional lives, too. Instead of nervously waiting for a job interview and obsessing about all the things that could go wrong, we can take a page out of Amy Cuddy's book and stand up, breathe deeply, and power-pose our way to lower cortisol, higher testosterone, and more confidence. Instead of using trendy leadership books and a new mission statement to fire up employees, we can follow ESADE's lead and use neurofeedback to heighten group coherence and prompt more productive strategy sessions.

But most of us, when challenged, will do none of these things. We'll think more, talk more, and stress more. We'll wait until *after* we feel better to go for that walk in the sun, rather than going for that walk *in order* to feel better. We'll wait until *after* we get that job offer to pump our fists and stand tall, instead of the other way around.

That's because, at first, reorienting from OS to UI can be downright disorienting. If I can change the "wallpaper of my mind" by deliberately shifting my neurophysiology—my breathing, my posture, my brainwaves, or any number of other interventions—what good are all those stories I've been telling myself? If I am not my thoughts, then who am I, really?

This idea, that our ego isn't the be-all and end-all, flourished in Asia for centuries before landing in California in the 1960's. Thoughts were illusions, the swamis and lamas maintained, and nirvana lay on the other side of ego death. But, for modern Americans, all those earnest (and sometimes addled) attempts to transcend the self didn't turn out to be that practical. To make sense of today's fast-paced world, we need our egos to navigate our relationships and responsibilities. We just don't need to use them like Maslow's hammer, turning everything around us into a psychological problem to beat on.

Instead, we can stay above our storytelling mind and simply monitor the knobs and levers of our neurobiology. And while this may seem far-fetched, top performers are already there. Tibetan monks can shut off their default mode network (or inner mind chatter) almost at will, SEAL snipers tune their brainwaves to the alpha frequency before locking on to targets, extreme athletes smooth out their heart rhythms right before dropping into a mountain or wave. They're deliberately doing an end run around their conscious minds. They're accessing more efficient and effective ways of being, and they're doing this exactly backward from how most of us have been taught.

Which brings us back to ecstasis. When we step beyond our conventional egos and experience the richness of altered states, it's essential to upgrade our software. Those monkey-suit personas we thought *were* us (until we suddenly realize they aren't) don't need to confine us or define us. "To diagnose . . . yourself while in the midst of action requires the ability to achieve some distance from those on-the-ground events," Harvard Business School professor Ron Heifetz maintains. "'Getting on the balcony' . . . [provides] the distanced perspective you need to see what is really happening."

And this is what moving from OS to UI delivers: a better view from the balcony. When we consistently see more of "what is really happening," we can liberate ourselves from the limitations of our psychology. We can put our egos to better use, using them to modulate our neurobiology and with it, our experience. We can train our brains to find our minds.

CHAPTER SIX

PHARMACOLOGY

Everybody Must Get Stoned

In 2012, in the waters off the southeastern coast of Africa, Emmy Award–winning wildlife photographer John Downer set up a series of hidden cameras in an attempt to nab footage of bottlenose dolphins in their natural habitat. He deployed underwater cameras disguised as squid, others hidden inside fish costumes, even some dressed up like sea turtles. All of this was done in the hope of capturing more relaxed behavior from the animals than would be possible with traditional filming techniques. And it worked. The dolphins in Downer's footage did appear more relaxed than normal—way, way more relaxed.

That's because the dolphins were high as kites on puffer fish. Caught on film for the very first time, a dolphin grabbed a puffer fish off the ocean floor, chewed on it for a little while, then passed it along to another dolphin in the pod. For a moment, it looked like the animals were playing a game of underwater catch, but it didn't take long for the terrorized puffer fish to release its primary defense mechanism, a yellowish cloud of deadly nerve toxin—which, in light of what happened next, seemed to be exactly the dolphins' plan.

While fatal in large doses, in small amounts puffer nerve toxin is intoxicating, inducing a potent shift of consciousness that produces a trancelike state in dolphins. After ingesting a hit, the animals in Downer's footage huddled in a tight pod, smiles on their faces, tails pointed toward the seafloor, snouts tickling the water's surface. "They were hanging around with their noses at the surface as if they were fascinated by their own reflection," Downer told the *International Business Times*. "It reminded us of that craze a few years ago when people started licking toads to get a buzz."

When the footage was released, it caused a bit of an uproar. Headlines like "Do Stoned Dolphins Give 'Puff Pass' a Whole New Meaning?" became popular, and YouTube viewings rose into the millions. But, really, stoned dolphins should have surprised no one.

Psychopharmacologists have spent the past few decades cataloguing the consciousness-altering techniques of animals in the wild and they have found plenty to document. Dogs lick toads for the buzz, horses go crazy for locoweed, goats gobble magic mushrooms, birds chew marijuana seeds, cats enjoy catnip, wallabies ravage poppy fields, reindeer indulge fly agaric mushrooms, baboons prefer iboga, sheep delight in hallucinogenic lichen, and elephants get drunk on fermented fruit (though they've also been known to raid breweries).

So commonplace is this behavior in animals that researchers have come to believe that, as UCLA psychopharmacologist Ronald K. Siegel pointed out in his book *Intoxication*, "drug seeking and drug taking are biologically normal behaviors. . . . In a sense, pursuit of intoxicating drugs [in animals] is the rule rather than the exception." This has led Siegel to a controversial conclusion: "The pursuit of intoxication with drugs is a primary motivational force in organisms."

So potent is the urge to get out of our heads that it functions as a "fourth drive," a behavior-shaping force as powerful as our first three drives—the desire for food, water, and sex. The bigger question is why. Intoxication, in animals as in humans, is not always the best strategy for survival. "The carcasses of drunken birds litter the highways," acknowledges Siegel. "Cats pay for their addiction to pleasure plants with brain damage. Cows poisoned with range weed may eventually die. . . . Disoriented monkeys ignore their young and wander from the safety of the troop. Human beings are no different."

But if mind-altering substances are so dangerous, why would any species take the risk? If the goal of evolution is survival and propagation, behaviors that threaten this mandate tend to get edited out over time. But the fact that drug use is as common in the jungles of the Amazon as it is on the streets of Los Angeles suggests that it serves a useful evolutionary purpose. Researchers have been pondering this for a while now, and have concluded that intoxication does play a powerful evolutionary role—"depatterning."

In nature, animals often get stuck in ruts, repeating the same actions over and over with diminishing returns. But interrupting this behavior is not easy. "The principle of conservation tends to rigidly preserve established schemes and patterns," writes Italian ethnobotanist Giorgio Samorini in his book *Animals and Psychedelics,* "but modification (the search for new pathways) requires a depatterning instrument . . . capable of opposing—at least at certain determined moments—the principle of conservation. It is my impression that drug-seeking and drug-taking behavior, on the part of both humans and animals, enjoys an intimate connection with . . . depatterning."

In more contemporary terms, both Siegel and Samorini have argued that animals consume psychoactive plants because they promote "lateral thinking," or problem solving through indirect and creative approaches. Lateral thinking involves big intuitive leaps between ideas. These are outside-the-box insights far more than iterative improvements, and much more difficult to achieve during normal waking consciousness. With our self forever standing guard over our ideas, crazy schemes and hare-brained notions tend to get filtered out long before they can become useful. But intoxication lessens those constraints.

And the evolutionary advantages of intoxication benefit all involved. In his book *The Botany of Desire,* Michael Pollan argues that coevolution—when two different species come together, often without knowing it, to advance each other's self-interest—also extends to humans and intoxicating plants. In return for helping mind-altering plants propagate and outcompete other species, these same plants have evolved even greater psychoactive properties for us to enjoy. "Plants," Pollan explained in a recent essay, "evolved to gratify our desires. . . . [In return], we give them more habitat and we carry their genes around the world. This is what I mean by the 'botany of desire.' Our desire . . . for intoxication, for changes in consciousness, [is] a powerful force in natural history."

But there are a couple of limitations that have long kept this co-evolutionary force in check. The first is location. Elephants are drunkards and not crackheads because coca grows in the Andes and not in Africa. Baboons never sample the mushrooms of the arctic tundra and have to rely on iboga for their kicks instead. Dolphins flirt with lethal poisoning because they can't score any hooch. Humans too have remained largely bound by geography. Until the arrival of global trade and travel, the substances we could use were the ones that grew around us.

The second limitation is culture. Anthropologists have discovered that as soon as a local intoxicant becomes enshrined in tradition, people grow suspicious of imports. "Most cultures," explains Pollan, "curiously, promote one plant for this purpose, or two, and condemn others. They fetishize one and they have taboos on others."

This explains why, when Franciscan priests arrived in Mexico and found the peyote cactus at the center of the local religion, they outlawed the plant, and enforced their own preference for sacramental wine (despite catastrophic consequences for the native populations, who were missing a key enzyme to metabolize ethanol). Conversely, in 1920's Prohibition America, growing apples—which could be fermented into hard cider—was against the law, but tinctures of opium and marijuana were readily available at the local pharmacy.

These are the constraints of the botany of desire: geography and culture. Together, they have prevented us from fully expressing that "fourth evolutionary drive," the irrepressible desire to seek nonordinary states of consciousness. And, since different chemical compounds unlock different states (and with them, unique and novel information streams), these restrictions have limited our access to the "repatterning" benefits of different types of cognition.

But pharmacology—and specifically, the branch of pharmacology that deals with psychoactive compounds—changes the rules of the game. It gives us access to more substances than ever before, and this provides us with more diverse data to consider. And perhaps no one played a bigger role in rewriting those rules than renegade chemist Alexander Shulgin.

The Johnny Appleseed of Psychedelics

Alexander Shulgin was called many names over the course of his career. *Wired* dubbed him "Professor X," while the *New York Times* preferred "Dr. Ecstasy." As he was a tall man with a shock of white hair and a thick beard, "Gandalf" was not uncommon. More formally, Shulgin has been described as "a genius biochemist," a "pioneering psychopharmacologist," and, according to the Drug Enforcement Administration (DEA), a "dangerous criminal." And his friends? His friends knew him as Sasha.

Sasha Shulgin was born in Berkeley, California, in 1925. Something of a boy wonder, he got a full scholarship to Harvard to study chemistry at age fifteen, but dropped out after a few semesters to join the Navy. After the end of World War II, he picked that interest back up, getting a Ph.D. from the University of California, Berkeley before taking a job as a senior researcher with Dow Chemical Company. It was there he made two discoveries that would shape the course of his life.

The first was Zectran, a biodegradable pesticide that turned into a highly profitable product for the company. The second was mescaline, a drug he tried for the first time while working for Dow. He was amazed that a tiny bit of white powder could produce such a big reaction. "I learned there were worlds inside of me," he said. And understanding those worlds became the central quest of his life.

"Sasha's interest," explains Johns Hopkins psychopharmacologist Roland Griffiths, was "exploring the nature and limits of the human condition through tweaking some of these [psychedelic] molecules to produce different and interesting kinds of effects. That's really, at its heart, what clinical pharmacology is about. We're studying the nature of the human organism."

After his blockbuster pesticide, Dow gave Shulgin the freedom to pursue his own interests. Not surprisingly, he chose to focus on psychedelics, starting with mescaline, modifying the drug one atom at a time, and testing the results on himself. As unorthodox as this may sound today, self-testing was the standard in psychopharmacology for most of the past century, and Shulgin took up this practice with vigor. He tried out every variation of mescaline he could create, then moved on to other compounds.

This all took place during the turbulence of the 1960s and Dow grew increasingly uncomfortable with holding patents on drugs the government was fighting to stamp out. When Shulgin's recipe for the potent amphetamine DOM was duplicated by underground chemists and sold to unsuspecting hippies—triggering a nationwide spike in emergency room visits—the company hit the breaking point. They decided it was time to part ways.

So Shulgin went into private practice, becoming an *extremely* independent contractor. After converting an old garden shed into his lab, he picked back up where he had left off: formulating and testing new psychedelics. Between 1966, when he first set up his backyard workshop, and his death in 2014, Shulgin became one of the more prolific psychonauts (an explorer of inner space) in history. He developed and tested, first on himself, next on his wife, Ann, then on the small circle of friends who served as his "research group," more than two hundred compounds. To put this accomplishment into perspective, in the 1950s there were about twenty known psychedelics, including LSD, psilocybin, and DMT. By the 1980's, Shulgin had expanded that list to more than two hundred.

Out of all these experiments, Shulgin is best known for resynthesizing MDMA, a compound first developed by German pharmaceutical company Merk in 1912, noticing it powerfully boosted empathy, and telling a few psychiatrist friends that it might have therapeutic potential. Those friends tried it out for themselves and were impressed with the results. They started using it with patients and those patients told their friends and pretty soon word hit the street of a love bomb in pill form and the Ecstasy craze was born.

Shulgin also cooked up stimulants, depressants, aphrodisiacs, and a bunch of stranger chemicals: drugs that slowed time, drugs that sped it up; drugs that produced powerful emotional reactions, others that deadened feeling completely. In addition, he developed the Shulgin Rating Scale, a six-level system going from Minus for "no result" to Plus 4 for "a rare and precious transcendental state," and took copious notes. "At 22 milligrams," he writes about the psychedelic 2C-T2, "a slow onset. It took an hour for a plus one, and almost another hour to get a +++. Very vivid fantasy images . . . some grey-yellow patterns a la psilocybin. Acute diarrhea at about the fourth hour but no other obvious physical problems. Erotic lovely (Shulgin's term for the quality of sex). Good material for unknown number of possible uses . . . Better try 20 mg next time."

Remarkably, Shulgin did all of this in plain sight of the law. In order to test these substances legally, he obtained a Schedule I license from the DEA (which gave him access to the most restricted class of drugs). Along the way, he developed a friendship with Bob Sager, head of the DEA's Western Laboratories, and began holding seminars for agents, supplying them with samples, and occasionally testifying in court. In 1988, he authored what became the field's definitive reference book, *Controlled Substances: Chemical and Legal Guide to Federal Drug Laws,* which won him several law enforcement awards.

But it was a pair of different books that came to define Shulgin's legacy. The first was *PiHKAL*, short for "Phenethylamines I Have Known and Loved," a reference to the class of psychedelics containing mescaline and 2C-B. Cowritten with his wife and published in 1991, *PiHKAL* was divided into two parts. Part One contained a fictionalized autobiography of the couple. Part Two was a detailed description of 179 psychedelics and included step-by-step instructions for synthesis, bioassays, dosages, duration, legal status, and commentary—that is, everything a would-be psychonaut needed for takeoff.

The second book, *TiHKAL,* came out in 1998, with the acronym standing for "Tryptamines I Have Known and Loved," and referring to drugs like LSD, DMT, and ibogaine. In this volume, the Shulgins included recipes for fifty-five more substances along with even more commentary. "Use them with care," they wrote, "and use them with respect as to the transformations they can achieve, and you have an extraordinary research tool. Go banging about with a psychedelic drug for a Saturday night turn-on, and you can get to a really bad place. . . ."

Not that their cautions prevented the inevitable banging about, or avoided the predictable consequences. Two years after they published *PIHKAL*, Richard Meyers, a spokesperson for the DEA's San Francisco office, told reporters: "It is our opinion that those books are pretty much cookbooks on how to make illegal drugs. Agents tell me that in clandestine labs they have raided, they have found copies of them." So they raided Shulgin's lab, fined him $25,000 for violating the terms of their agreement, and yanked his Schedule I license.

While a variety of government agencies continued to surveil Shulgin for the rest of his life, he was never charged with a crime. Nor, as Rick Doblin, founder of the psychedelic research nonprofit MAPS, explains, did he ever regret his decision.

"Sasha," Doblin says, "was fiercely in favor of personal liberty. He felt these kinds of consciousness-expanding experiences were crucial to the world's spiritual and emotional development. His decision to share his research came from a real fear that he would die with this enormous body of knowledge trapped inside him. Even before *PiHKAL,* Sasha had that open-source impulse. He gave away information to anyone who asked—it didn't matter if they were DEA agents or underground psychedelic chemists. But after publication, when the crackdown came, it was like locking the barn door after the horses had already gone. The research was out there and Sasha knew, even if the contemporary climate was hostile to these substances, sometime in the future things would change and his work would be very useful."

By publishing his psychedelic cookbooks, Shulgin bypassed the geographical and cultural limits of the botany of desire. In open-sourcing these recipes, he distributed hundreds of tools for investigating consciousness and changed countless lives. "Everybody knows who the Shulgins are," wrote Teafaerie, a close friend of the couple, in her widely circulated essay: "No Retirement Plan for Wizards." "It's pretty much impossible to overstate their collective contribution to psychedelic culture, and indeed to the very fabric of human society at large. They not only brought us most of our favorite alphabetamines, they tested them on themselves and published their extensive notes so the rest of us could benefit from their groundbreaking discoveries. Sasha is the greatest psychopharmacologist who ever lived. Ann is a pioneer in the field of empathogen-assisted therapy. Their love story has inspired millions of people. And that's just for starters."

This Is Your Brain on Drugs

While Shulgin definitely got things started, his impact was primarily felt on the edges of society—in law enforcement and the counterculture. But it's next-generation psychedelic researchers like Robin Carhart-Harris who have brought that impact directly into the mainstream.

Carhart-Harris didn't start out interested in mind-altering substances. In the beginning, when completing a master's in psychoanalysis at Brunel University in England, it was the unconscious that caught his attention. "Here was a part of our mind that seemed to govern so much of our behavior," he explains, "yet it was incredibly difficult to study. I was in a seminar where the class leader rattled off all the different methods we use to access it—free association, dream analysis, hypnosis, bungled actions, slips of the tongue. None were very good. Except for dreaming, they're all indirect approaches. And dreaming takes place when we're asleep, so all we can get is after-the-fact reports. If we were going to make any headway on this problem, we had to find a better way to explore the unconscious."

In his hunt for that better way, Carhart-Harris picked up psychologist Stanislav Grof's classic book, *Realms of the Human Unconscious: Observations from LSD Research*. One of Grof's main arguments was that during psychedelic states, our ego defenses are so diminished that we gain nearly direct access to the unconscious. That's when it clicked. With tools like fMRI, Carhart-Harris could exploit this access—he could take pictures of the unconscious in real time.

After finishing his master's, he switched careers and joined David Nutt's lab (the psychopharmacologist we met in Chapter 3) at the University of Bristol, learning the basics of neuroimaging doing sleep research and making an initial foray into psychedelics by imaging MDMA. In 2009, he became head of psychedelic research at Imperial College London and became the second person in history to use fMRI to explore the neurological impact of psilocybin. And the very first to explore LSD.

These were important milestones. "This is to neuroscience what the Higgs boson was to particle physics," David Nutt told the *Guardian*. "We didn't know how these profound effects were produced. It was too difficult to do. Scientists were either scared or couldn't be bothered to overcome the enormous hurdles to get it done."

More than anything else, what those first-ever imaging studies revealed was the middle ground that Sasha Shulgin never had the tools to map. Shulgin gave us a wider assortment of chemicals to assay and subjective reports about what happened when we did. Carhart-Harris closed this gap. He showed us what was going on in the brain, illuminating the neurological mechanisms that lay beneath Shulgin's subjective reports.

And these newly discovered mechanisms shed more light on two of the fundamental characteristics of ecstasis: selflessness and richness. Earlier in the book, we explored how the deactivation of key parts of the brain, what's called *transient hypofrontality*, is largely responsible for selflessness. Carhart-Harris extended this work, helping to determine exactly which parts are involved in that process. "A lot of the earlier imaging work on altered states gave us static pictures of the brain. So we made correlations: when we're on LSD, this region deactivates; when we're meditating, that region deactivates. But the technology has improved and we can now take dynamic pictures. This is how we know that the vanishing of self is not really about specific regions deactivating. It's bigger than that. It's more like whole networks disintegrating."

One of the most important networks to disintegrate is the default mode network. Responsible for mind-wandering and daydreaming, this network is active when we're awake but not focused on a task. It's the source of a lot of our mind chatter, and with it, a lot of our unhappiness. But, like many of the brain's systems, the default mode network is fragile. A little trouble in a couple of nodes is all it takes to knock it offline. "Early psychologists used terms like '*ego disintegration*' to describe the effects of an altered state," says Carhart-Harris. "They were more correct than they knew. The ego is really just a network, and things like psychedelics, flow, and meditation compromise those connections. They literally dis-integrate the network."

The other important discovery made by Carhart-Harris and his team involved the birth of new networks. The scans revealed that psychedelics created highly synchronized connections between far-flung areas of the brain, the kinds of linkages we don't normally make. So when researchers like James Fadiman discovered that psychedelics could enhance creative problem solving—these far-flung connections were the reason why. Or, as Carhart-Harris explains, "What we've done in this research is begin to identify the biological basis of the reported mind expansion associated with psychedelic drugs."

Carhart-Harris set out to take real-time pictures of the unconscious and when he did, he saw the unconscious actively hunting for new ideas. It's a discovery that helps to legitimize these substances as performance-enhancing tools for solving wicked problems. And it's one that Carhart-Harris feels couldn't come at a better time.

"A lot of people have been pointing out that the modern world is in crisis. I don't know if I agree with the most pessimistic of those assessments, but I do know it takes significant cognitive flexibility to solve complex problems. So I do think all this research is timely. It's making us a little less afraid of a powerful problem-solving tool. Going forward, I have a hunch that's going to matter."

The Hyperspace Lexicon

On September 22, 1823, a seventeen-year-old farm boy from Manchester, New York, named Joseph Smith had a strange dream about an angel named Moroni. The angel told him of a treasure buried on a hilltop behind his house. Upon awakening, Smith climbed that hill and, just shy of the peak, unearthed a gold-leafed book. Bound together with three D-shaped rings and written in strange hieroglyphics he later described as "reformed Egyptian," it contained a prophecy that would alter the course of U.S. history.

The book told of a lost tribe of Israelites who had sailed to North America in 600 BCE. It recounted the story of a prophet named Mormon and the second coming of Jesus Christ. If true, these ideas would turn nearly two thousand years of Christian orthodoxy on its head.

But there was one small problem: proof. The angel didn't let Smith bring the golden tablets down "Mormon Hill." In fact, by the time Smith translated the revelation and published it a few years later as the *Book of Mormon*, the golden plates were nowhere to be found. The angel, Smith reported, had taken them back for good.

While many of Smith's contemporaries doubted his story, and subsequent scholars found no evidence of a "reformed Egyptian" culture in North America, to the faithful, Smith's account stands as gospel. So compelling was his epiphany that it inspired one of the most successful religions in American history. The Church of Jesus Christ of Latter-day Saints transformed the barren deserts of Utah into a garden theocracy, constructing massive temples and a global mission network that impacts millions to this day.

And Joseph Smith was by no means the first person to have a prophetic vision that then birthed a religion. Moses fathered three of the world's largest traditions—Judaism, Islam, and Christianity—when he came down from Mount Sinai with two stone tablets written by "the finger of God." But this time as well, the problem was proof.

When Moses returned to camp, he found the Israelites worshipping a golden calf. Enraged at their idolatry, he smashed the newly forged Ten Commandments into shards. It was the first time in recorded Western history that God had directly communicated with humanity, and the evidence was destroyed almost as soon as it had been created. The chastened Hebrews had to take Moses's word for it.

Until recently, those sorts of visions were always unverifiable. Access to ecstasis was rare. Understanding its mechanisms and meaning rarer still. When someone did find a hotline to God, the experience was usually a one-off: unrepeatable and impossible to validate. The only way to gauge the truth revealed in these states was to assess the conviction of the original visionary or listen to the retroactive stories of their followers.

But Sasha Shulgin and those researchers who have come after him have changed our relationship to revelation. Because, say what you will about pharmacology being a cheat code to the mystical, there's no question it works. "[S]ome people can reach transcendent states through meditation or similar trance-inducing techniques," New York University neurologist Oliver Sacks once explained, "[b]ut drugs offer a shortcut; they promise transcendence on demand." No climbing a mountaintop and waiting for lightning to strike. No sitting on a cushion wondering if nirvana will ever arrive. Pop the pill, take the hit, and pretty soon you're undeniably someplace else. This ease of access means more people can visit these states more often and collect more data. As a result, they no longer have to treat one person's epiphany as written-in-stone Truth.

Not long after the publication of *PiHKAL* and *TiHKAL*, online message boards and forums began popping up to provide clandestine recipes for kitchen chemists and detailed maps for explorers of inner space. Referring to Erowid, one of the largest and best-regarded of these online repositories, author Erik Davis writes: "By far the most entertaining vault contains thousands of 'experience reports' logged by psychonauts flying high (and taking notes) on exotic cacti, prescription pharmaceuticals, and newfangled phenethylamines. . . . At once formulaic and bizarre, these reports provide details . . . largely lacking in the hazy trip tales of yore."

This open-source approach to pharmacology has given us a

way to fact-check ecstatic inspiration, moving us from the "one to many" route—à la Moses and Joseph Smith—to a "many to many" model. Rather than having to take anyone's word for what happens out there, explorers can now repeat the original experiments and see for themselves. It's a development that upends the closely guarded and often reactionary world of revealed truth and religious insight.

Take, for example, psychiatrist Rick Strassman's research at the University of New Mexico. In the early 1990s, Strassman was searching for a naturally occurring substance in the human body that could prompt mystical states, which he hoped might provide a way to explain the epiphanies of Moses and so many other historical prophets. Initially, he turned his attention to melatonin but, disappointed with the results, soon decided to focus on its cousin DMT (dimethyltryptamine). DMT made sense as a candidate. It occurs naturally in the human body yet when vaporized or injected, becomes a powerful psychedelic.

Strassman was shocked by how powerful it was: "I had over twenty years in experience and training and study in [a] Zen monastery, and I was expecting those kinds of enlightenment experiences from . . . DMT." But Strassman's study group didn't experience anything that could comfortably be described as Buddhist. More than 50 percent of his research subjects blasted off to distant galaxies, had hair-raising encounters with multidimensional entities, and came back swearing that those experiences felt "as real, or in many cases, more real than waking life."

When Strassman published his findings in the Buddhist review *Tricycle,* there was a considerable backlash. He was roundly criticized by readers and disowned by his meditation community for suggesting there might be something *else* going on in mystical states besides Zen oneness. In 1995, it all became too much. He shut the project down, sent the remaining chemicals back to the DEA, and retired to the mountains of Taos, New Mexico, to knit alpaca sweaters.

But the precedent Strassman set proved compelling. In the wake of his study, unsanctioned use of DMT and its more potent cousin, 5-MeO-DMT, skyrocketed. An experience too strange for all but the most ardent sixties' trippers became the first truly Digital Age psychedelic—knowingly referenced in DJ samples, discussed at length in podcasts, and inspiring online forums all its own.

One forum in particular, the Hyperspace Lexicon, reflects a collective effort to codify and make sense of the utterly novel landscape of DMT (which aficionados refer to as "hyperspace"). The Lexicon is packed with neologisms that would have made James Joyce proud. Among many others, there's "*lumenorgastic,*" for the orgasmic experience of white light; "*mangotanglement,*" referring to the brightly colored fractal building blocks of DMT reality; and "*ontoseismic,*" for the utter shattering of your worldview after glimpsing the DMT universe. But tucked beneath the creative nomenclature, the Hyperspace Lexicon reflects a watershed in how we relate to ecstatic revelation.

For a concept to make it into the Lexicon, it can't have come from anyone's singular experience. Rather, the new idea has to resonate with a critical mass of the community. And even then, it's taken with a grain of salt.

Consider the first entry, under "A," for "Akashic Book," an imagined but seemingly real book that, as the Lexicon says, "contains deep knowledge in a language you can't read but do understand. It is so profound . . . it contains exactly the right wisdom for you at this moment." Whether Moses and Joseph Smith perceived that same Akashic Book as stone tablets or golden plates is impossible to prove (though the similarities to these reports are notable). However, unlike the lessons of these prophets, no one is being asked to take insights gleaned from that book at face value. While the Lexicon specifies that the knowledge it contains is profound, even omniscient, it then qualifies that with "for you at this moment."

Another example of the provisional nature of the Lexicon is the phrase "End of the Line." Defined as: "When having a DMT experience and you feel as if you have reached the Absolute Point. The Alpha and Omega of the universe and your entire existence. Then you are having a 'The End of the Line' type breakthrough. It may very well not be true at all, but for you, subjectively, it feels *as if*. So, even though an "End of the Line" experience might have birthed hellfire zealots in earlier days, today we have anonymous explorers bracketing the certainty of their experience with phrases like "you feel as if" and "it may very well not be true at all."

If you put this all together, what seems to be emerging in the aftermath of Shulgin, Carhart-Harris, and Strassman is a kind of "agnostic Gnosticism," an experience of the infinite rooted in the certainty that all interpretations are personal, provisional, and partial. As a result, no one can claim their particular vision of the divine as correct, if there are thousands of other "visions" with which to compare it. And anyone who does try to claim the spotlight? Even a few decades ago, they could have started a cult. These days, they'll just get trolled online, then ignored.

And what that does is leave more room for open experimentation. It disempowers anyone tempted to escalate his position and privilege, and empowers everyone else to make sense of their own experiences.

When the physicist Enrico Fermi famously guessed the number of piano tuners in Chicago, or the number of stars in our galaxy, he did so by applying provisional estimates to impossibly large problems. And while never exact, his guesses often landed within an order of magnitude of the actual number—enough, in other words, to act upon.

Today we're following Fermi's lead, applying the power of Big Data to approximate answers to the Big Questions. One or two data points like a Moses or a Joseph Smith can't ever make a trend, but what about a thousand data points? A hundred thousand? A picture is starting to emerge of the worlds inside us. And while it's no less strange, it is arguably a good deal more accurate than the singular epiphanies that have come before.

The Molecules of Desire

In 2010, chemist Lee Cronin saw a demonstration of 3D printing while attending a conference in London. The technology caught his attention. At his own lab, he was always having to fabricate equipment. So, he wondered, could 3D printers solve this problem?

Cronin returned to the University of Glasgow, where he was a professor, and organized a feasibility workshop. It didn't take long to discover that bathroom sealant—the kind available in any hardware store—could be used as raw material, letting him print test tubes and beakers in any size or shape. He also found that the printer could be used to create simple molecules. By filling it with "chemical inks"—the reagents from which molecules are built—Cronin prototyped a rudimentary "chem-printer."

Since then, Cronin has moved past formulating simple molecules and into more complicated compounds. His short term goal is to figure out how to print over-the-counter medicines like ibuprofen. His long-term goal is to create a set of universal inks capable of making any drug from scratch. "Almost all drugs are made from simple molecules like carbon, hydrogen, and oxygen," Cronin explains, "as well as easily available materials like paraffin and vegetable oil. With a 3D printer it should be possible that with a relatively small number of inks you could actually make any organic molecule."

This discovery would make prescription drugs downloadable, allowing anyone anywhere access to the medicines they need. But it will do far more than that. "Of course," says Cronin, "this printer will also lead to new frameworks for discovery and increase the possibilities of new kinds of mind-altering drugs."

While all manner of psychoactive plants are available online, allowing the adventurous to distill potent psychedelics with little more than a Crock-Pot, some Mason jars and a turkey baster, the DEA and INTERPOL can still shut down these gray market suppliers. But Cronin's 3D drug printer renders that kind of oversight almost impossible. After all, how can you regulate access to controlled substances when the raw materials have become as pedestrian as paraffin wax and vegetable oil?

What's more, 3D printers have user-friendly interfaces, so all that's required to work one is the ability to point and click a mouse. And they're inexpensive, so using them doesn't require big grants or costly labs. Their combination of simplicity and affordability puts every step of the chemical formulation process within reach of anyone with an internet connection and a power outlet. So, for those interested in taking one of the formulas from Shulgin's cookbooks and modifying it to create the next great alphabetamine? Simply hit print.

And Cronin's 3D chem printer isn't the only development reshaping pharmacology. By treating the four letters of the genetic alphabet like the ones and zeros of computer code, synthetic biology gives us the ability to program living cells with the same ease that we program computers. And once the code looks right? Send it to a DNA synthesizer and within a few days you'll receive a freeze-dried vial in the mail, filled with your genetic creation.

Not surprisingly, synthetic biologists have already figured out how to use this process to produce psychoactive substances. In August 2014, researchers at Stanford announced they had genetically engineered yeast to produce the painkiller hydrocodone. Conventionally farmed, poppy plants take about a year to grow enough opium to make this drug, but this new yeast can do the work in days. The Canadian company, HyaSynth Bio, meanwhile, is engineering a different yeast to produce THC and cannabidiol, two of the active ingredients in marijuana.

"This is really just the beginning," explains Autodesk distinguished researcher and synthetic biologist Andrew Hessel. "Pretty much any substance made by a plant, tree or mushroom, including all the neuroactive substances, is within reach of synthetic biology. We're not there yet, but within a decade or so this one technology should be able to tickle all the same receptor sites in the brain that mind-altering substances impact."

Author Michael Pollan argued that the botany of desire—the idea that intoxicating plants shape us as much as we shape them—has played an unsung but important role in the evolution of culture. But, no matter how influential these plants may have been, historically our ability to use them has been held in check. If the plant didn't grow nearby, if it wasn't on the list of socially sanctioned substances, potential explorers were out of luck.

Pharmacology is such a potent force for ecstasis because it changes the nature of this game. With chemical cookbooks, rigorous neuroscience, crowdsourced lexicons, and now, democratized means of production, we're freed from the geographic and cultural limitations we inherited. By giving us access to not just the botany of desire but the *molecules* of desire, we can continue to shape these compounds even as they inevitably shape us. It's coevolution compressed from millennia into minutes.

CHAPTER SEVEN

TECHNOLOGY

Dean's Dark Secret

The why was never in question. What happened? How it happened? Those answers remain unclear. But the why? For Dean Potter—it was never in doubt.

It was May 16, 2015, in Yosemite Valley, California, a nice spring evening, right on the edge of dusk. Potter, a record-breaking rock climber, slackline walker, and wingsuit flier, got ready for the evening's adventure. He was 3,500 feet above the valley floor, standing on the summit of Taft Point. Alongside his friend and fellow flier Graham Hunt—considered one of the best young pilots around—their goal was to leap off the edge, zip across the canyon below, and sail through a V-shaped notch in a neighboring ridge, above an ominously named cliff, Lost Brother.

Dean Potter played an important role in the writing of *Rise of Superman*. He was a good friend of the Flow Genome Project, a member of our advisory board, and as big-hearted and thoughtful as any professional athlete we've met. In 2013, when we were filming the *Rise of Superman* video series, Dean told the story of how he nearly died while BASE-jumping into a deep cave in Mexico. He finished tellingly: "This year, twenty-something wingsuiters have lost their lives. Dying's not worth it. I've been struggling with that a lot. I don't want to be that guy who got lucky. And I've been that guy who got lucky for a lot of years. I want to be that guy that's such a wizard of strategy and knows myself and am comfortable enough to say, 'Na-ah, I'm not going. I want to live.'"

But, that early evening in Yosemite, he went anyway.

Graham and Dean launched off Taft Point. Forty seconds later, their flights were over. Both men came into the notch low, possibly because the colder, denser winds that arrived with the setting sun had cost them altitude. Potter never wavered, but Hunt—as far as anyone can tell—jerked left, then swerved right, putting him on a diagonal path and directly into the canyon's far wall. Potter made it through the notch, but didn't have the height he needed and crashed into the rocks on the other side. Both men died on impact.

And to this day, the details of the accident remain mysteries. No one knows *what* caused Hunt to swerve; no one knows exactly *how* Potter lost so much altitude. But the *why* was never in question.

"Look," Potter once explained, "I know the dark secret. I know my options. I can sit on a cushion and meditate for two hours and maybe I get a glimpse of something interesting—and maybe it lasts two seconds—but I put on a wingsuit and leap off a cliff and it's instantaneous: Whammo, there I am, in an alternate universe that lasts for hours."

And for flow junkies who get their fix through action sports, this has always been the dark secret. Ecstasis only arises when attention is fully focused in the present moment. In meditation, for example, the reason you follow your breath is to ride its rhythm right into the now. Psychedelics overwhelm the senses with data, throwing so much information at us per second that paying attention to anything else becomes impossible. And for action and adventure athletes seeking flow, risk serves this same function.

"When a man knows he is to be hanged in the morning," Samuel Johnson once remarked, "it concentrates his mind wonderfully."

By 2015, wingsuiting was providing exactly that kind of dangerous focus. "I start to shiver and wonder if what we're doing is right," Dean wrote in an essay just a month before his death. "Wingsuit BASE-jumping feels safe to me, but [so many] fliers have lost their lives this year alone. There must be some flaw in our system, a lethal secret beyond my comprehension."

The lethal flaw is that, for many, using high-risk sport to explore ecstasis is so compelling and rewarding that it becomes an experience worth dying for. Steph Davis, Potter's ex-wife and a professional climber and wingsuiter herself, has lost two husbands to the sport, yet she keeps flying. The siren song of "hours in an alternate universe" that Dean sought has continued to beckon to pilots convinced they can dodge the rocks.

But for the rest us? Those with lives and wives and things that matter? Are we shut out of these "alternate universes"? Do we have to make an impossible choice between dedicating decades to practice or accepting intolerable risks to get there faster?

Thanks to inventors like skydiver Alan Metni, the answer, increasingly, looks like "no." Metni began his professional life as a lawyer at Vinson & Elkins, a global firm that counts senators, U.S. attorneys general, and Fortune 100 CEOs as alumni. But the legal life didn't do it for him, so he chucked it for his true passion: jumping out of airplanes. He pitched a tent at a local airport and began training relentlessly, logging more than ten thousand jumps and earning three U.S. national championships and a world championship in formation flying along the way.

But Metni wasn't satisfied. He wanted to find a way to train even harder, so he started tinkering with giant fans and wind tunnels. By the early 2000's, he'd perfected an indoor skydiving experience indistinguishable from true free fall. Suddenly, competitive teams could log hundreds of hours training together in absolute safety. With this one innovation, the standard of excellence at the elite level changed nearly overnight. Even SEAL Team Six came to work with Metni. Not to learn how to jump out of airplanes—they had that part down cold—but to train teamwork, group flow, and the secret to "flipping the switch" while falling through space together.

"Around the world," Metni said, "it doesn't matter what culture, language, or faith, everyone has the same dream: to fly." So he built a company, iFly, and set out to fulfill that dream, one wind tunnel at a time.

Today, iFly is in fourteen countries with over fifty-four tunnels and revenue nudging ten figures. Thousands and thousands of people who would never have considered jumping out of a perfectly good airplane or leaping off a cliff in a wingsuit have realized that dream, and done so safely. By taking out the risk, iFly has taken a sport once reserved for daredevils and made it accessible to everyone—ages three and up.

And skydiving isn't the only high-risk pursuit that has undergone a revolution in accessibility. Across the action sports industry, advancements in technology are providing safer and easier entry into the zone than ever before. Powder skiing, with its utterly magical sensation of floating down a mountain, used to be the rarefied domain of top athletes. Today, extra-fat skis make that float available to anyone who can link two turns together. Mountain bikes, which once offered bone-rattling descents to all but the best riders, now have supple front and rear suspensions, oversize balloon tires, and an ability to roll over the most daunting terrain. Even kitesurfing—best known on the internet for its "kitemare" footage of people getting dragged by giant sails across highways—has mellowed. Better safety gear lets newcomers find the balance between wind and waves with a fraction of the exposure and learning time.

This trend, of technological innovation providing wider and safer access to altered states, isn't limited to adventure sports. As we'll see in the rest of this chapter, it's showing up across many disciplines, allowing more people than ever before to sample what these experiences have to offer. We're shedding some light on Dean's dark secret, sparing many of us the stark trade-offs that he and so many other pioneers were forced to make. Technology is bringing ecstasis to the masses, allowing us to taste it all, without having to risk it all.

Things That Go Boom in the Night

For nearly fifty years, Tony Andrews has been messing with your mind. If you go out at night, to concerts or clubs, he's been there, hidden from sight perhaps, but tinkering with your head. If you've ever visited the Grand Ole Opry or the Royal Albert Hall or, for that matter, Space—the iconic Ibiza club voted best in the world—he was in that mix too, fine-tuning your brain with incredible precision.

Andrews's trade is audio. Technically, he's a sound engineer who makes loudspeakers. He's the cofounder of the British speaker company Funktion-One, and while the name may not be familiar, it represents a half century of sonic innovation and the quest for what Andrews has come to call "the audio moment."

The audio moment is an instant of total absorption. "It's the point," explained Andrews in a recent talk, "when you get really involved in the music. [When] you suddenly realize that you've been somewhat transported to another place. . . . When you find yourself experiencing more of yourself than you realized was there in the first place."

Andrews's original audio moment took place when he was nine years old. His mother had brought home a copy of Elvis Presley's "Hound Dog." When she put it on, immediately, Andrews felt the shift. "I was feeling all this electricity in my body, and I didn't understand what it was, but I knew it felt good—and I've been pursuing it ever since."

By the time he was sixteen, Andrews was tinkering, building his own tweeters and cabinets and testing out new ideas. One day, for instance, he placed a loudspeaker in the corner of their garage, with the speaker facing inward—so the sound shot straight into the intersection of the wall and floor. The high notes and the midrange vanished, but the bass was booming. So Tony designed a speaker surrounded by three walls, which looked a little like the bell of a trumpet. It was the discovery of "horn-loaded bass," an invention that led to a significant improvement in the acoustic quality of live concerts and, by extension, the number of transformative audio moments concertgoers could experience.

Andrews also realized that while a decent home stereo, what used to be called "hi-fi," had more than enough sound quality to shift consciousness, there were only so many bodies you could pack into a living room. If Andrews wanted to have a real impact, he needed to figure out how to bring that same high-quality sound to more people, to build a system capable of reproducing the audio moment at a stadium level. "Then," he says, "in my simplistic terms, people could have a common place to expand their minds."

And when we had the opportunity to stand in front of a Funktion-One sound stack at a recent concert, we got a clear sense of how far Andrew's had come in perfecting his "audio moment." The experience was less like listening to conventional music and more like getting pressure-washed with a sonic firehose. The bass hit our chests like a concussion grenade; the high notes made the hair on our necks stand up. And throughout the experience, there was no room for thinking at all.

Of course, Andrews is by no means the first to connect the power of music to its ability to shift consciousness. In recent years, scientists have found that a great many of the world's oldest religious sites have peculiar acoustic properties. While studying the Arcy-sur-Cure caves in France, University of Paris music ethnographer Iegor Reznikoff discovered that the largest collection of Neolithic paintings are found more than a kilometer deep. They'd been intentionally located at the most acoustically interesting spots in the cave: the parts with the most resonance.

"Reznikoff's theory," writes author Steven Johnson in *How We Got to Now,* "is that Neanderthal communities gathered beside the images they had painted, and they chanted or sang in some kind of shamanic ritual, using the reverberations of the cave to magically widen the sound of their voices."

When musicologists traced this trend through history, they found that what began in prehistoric caves also showed up in medieval churches. In Greece, churches were designed with narrow walls to produce a "slap echo," sort of a threefold sound bounce meant to represent the flapping of an angel's wings. In France, the Gothic arches of Notre Dame and Chartres cathedrals act as giant subwoofers for the pipe organs. For thousands of years, we have been engineering sound to shift consciousness.

"[I]n all societies," explains neurologist Oliver Sacks in the journal *Brain,* "a primary function of music is collective and communal, to bring and bind people together. . . . One of the most dramatic effects of music's power is the induction of trance states. . . . . Trance—ecstatic singing and dancing, wild movements and cries, perhaps rhythmic rocking, or catatonia-like rigidity or immobility . . . [is a] profoundly altered state; and whilst it can be achieved by a single individual, it often seems to be facilitated in a communal group."

Scientists working in the burgeoning field of neuro-musicology have begun using high-powered imaging to decode these effects. When listening to music, brainwaves move from the high-beta of normal waking consciousness down into the meditative (and trance-inducing) ranges of alpha and theta. At the same time, levels of stress hormones like norepinephrine and cortisol drop, while social bonding and reward chemicals like dopamine, endorphins, serotonin, and oxytocin spike. Add in entrainment—where people's brains synch to both the beat and to the brains of those around them—and you've got a potent combination for communitas.

In a recent study, Apple and the speaker manufacturer Sonos took a deeper look at music's power to connect. To track how much music people listened to at home (on average, four and a half hours a day) and what happened while they listened, they rigged thirty homes with Sonos speakers, Apple watches, Nest cams and iBeacons. When tunes were playing, the distance between house-mates decreased by 12 percent, while chances of cooking together increased by 33 percent, laughing together by 15 percent, inviting other people over by 85 percent, saying "I love you" by 18 percent, and, most tellingly, having sex by 37 percent.

This also explains why the honor of opening the main stage at Portugal's 2014 Boom Festival—essentially Europe's version of Burning Man—didn't go to a movie star or rock star or famous DJ, and instead fell to Tony Andrews, a loudspeaker manufacturer. "About forty-five years ago," Andrews told that audience, "I came to a place where I realized that . . . big sound can facilitate our communal mind moving to a place of . . . unity. I believe this is a really necessary step for humanity to take."

Some 26,000 people attended Boom. As the entire festival grounds had been—thanks to Andrews and the Funktion-One team—wired for big sound, nearly every one of them could experience that communal mind shift together. "What we're building in the Dance Temple," explained one of its designers, "is a piece of tech to disintegrate peoples' egos en masse."

And this is the biggest change that new technology has afforded. Drums and voices only carry so far. There are only so many people you can cram into a cave or a church. But 26,000 people is somewhere between a large town and a small city. Electronic festivals in Las Vegas and Miami pack in over a quarter of a million. Never before have so many people been able to come together and follow the beat right out of their minds.

The Digital Shaman

One of Tony Andrews's collaborators at Boom Festival was Android Jones, an unassuming visual artist with a knack for arresting images. A veteran of George Lucas's Industrial Light & Magic, as well as the first non-Japanese illustrator at the legendary game maker Nintendo, Jones has established himself as one of the most prolific creators in the visionary art scene.

Combining a classical fine arts education with the power of digital software, Jones creates images that defy easy categorization: archetypal deities overlaid with fractal geometries, cosmic lovers projected across giant galaxies, and ornate masks stretched across crystalline hillsides. At Boom fest, he "live-painted" his images onto huge screens as dancers writhed and stomped: creating an animated stained glass window for the Church of Trance. In the same way that Funktion-One is fine-tuning soundscapes, Jones is extending the impact of visual art.

Since 2011, he has partnered with Obscura Digital to project his images onto iconic public buildings around the world. On the Sydney Opera House, for instance, he live-painted during a performance by the YouTube Symphony Orchestra. He's also transformed the Empire State Building, the United Nations headquarters, and most recently, St. Peter's Basilica at the Vatican. "Digital projection mapping has created unparalleled possibilities," Jones says. "It takes art out of museums and brings it directly to people going about their daily lives. The scale of access it provides outpaces anything available in the past."

In the past, sacred art was viewable only in sacred places. If you wanted to look at prehistoric bulls, you'd have to descend into the caves of Lascaux. Only the faithful making the pilgrimage to Rome could view Michelangelo's Sistine Chapel *in situ*. The fact that Jones's artwork lives in a computer file and not stretched on canvas offers many more opportunities for display. When the goddess Kali is forty stories high on the side of the Empire State Building or a god's-eye montage of the planet's history is projected onto the side of the Vatican—people can encounter the sacred in the midst of the mundane.

And visionary art is not only becoming much more accessible;

with the internet, it's becoming interactive as well. "It's impossible to separate the feedback I get, online or in person, from the creative process itself," says Jones. "It's no longer just a one-way experience. I get all this nonverbal feedback—likes, shares, etc.—that directly informs the work."

The feedback isn't all "likes" and retweets, either. Android frequently reinterprets religious symbols and sacred iconography, riffing on thousands of years of culturally loaded imagery. For this, Jones has been dubbed a "pop shaman," but not everyone welcomes the impact his high-fidelity visuals can have. Beginning with his Empire State and UN installations, and then spiking after the exhibit at the Vatican, Android has spawned a lively cottage industry of YouTube conspiracies. He stands accused of everything from smuggling pagan idolatry into the Church's holy sites to programming unsuspecting minds with images from the Illuminati. One particularly cautious whistle-blower considered Jones's art so psychoactive that he wrote "I cannot even continue to view these images, even for the purposes of alerting others as to their true intent."

"At first I was laughing it off," explains Jones, "how funny people are and desperate to make meaning—making random speculations. But it isn't just speculation. There's something happening [to viewers] on an objective level."

Research conducted at Stanford supports Jones's hunch. A 2012 study found that encounters with perceptual vastness, be it the endless spiral of galaxies in the night sky or Jones's larger-than-life projections, triggers a self-negating, time-dilating sense of awe. And this happens automatically—which means an encounter with Jones's projections could be enough to drive subjects into a deeply altered state, willingly or not.

In March 2015, Jones upped the impact of his work even further. Joining forces with a group of Russian monks living in Thailand (who, improbably, happened to be tech whizzes), he moved from projecting his art onto flat 2-D surfaces into fully immersive 3D experiences. The coder-monks crafted Jones's original media files into a string of modular scenes. When stitched together and projected onto the smooth 360-degree canvas of an enclosed geodesic dome, the screen seems to extend far beyond the physical space. Rather than looking up at the curved vault of the Sistine Chapel to view a singular mural, you can now walk through an immersive and infinitely changeable landscape, a dream state filled with gods and demons, stardust and galaxies, and anything else Jones can dream up.

"After working in this 3D immersive space," he admits, "it's really challenging to go back to creating images that are on a rectangle hanging on a wall. I never realized how limiting the frames were." And he's not the only one exploring these possibilities.

Virtual and augmented reality companies like Oculus and Magic Leap are attracting outsize media and investor attention as everyone rushes into the landscape beyond frames and screens. They are early indicators of a new way of consuming content that increasingly blurs the boundaries between what is real and what is simulated. But perhaps more than any other artist, Jones is taking advantage of this technology to knock people out of their normal frames of reference, and give them a glimpse of ecstasis.

His most recent project, appropriately named *MicroDoseVR*, is an immersive VR game offering an atom's-eye tour through many of Shulgin's alphabetamine compounds. Zooming through that digital world, surrounded by deep trance music and the actual "molecules of desire," the simulation is more than enough to knock you out of regular awareness. "That's probably the real value of these experiences," explains Jones. "They take us out of our conditioned world. They open a realm of everything else we might never have experienced and only dreamed of. You think you know where the boundaries are, but you see this stuff and think, if this thing I'm looking at is possible, what else might be possible?"

Enlightenment Engineering

In 2011, Mikey Siegel, an MIT- and NASA-trained roboticist, was living in Silicon Valley and working his dream job. "It was engineering heaven. I got to build robots, design systems, code software, create my own experiments, and they paid me really well. I had a long list of everything I thought I wanted and everything on that list was ticked off." But "everything" wasn't cutting it.

Siegel felt anxious and unfulfilled, like his life had little substance, like, as he explains, "his soul was bankrupt." So he did what the unfulfilled have often done: went on a quest. He trekked through the jungles of South America and visited ashrams in India. His perspective shifted, but the vision he'd been questing for never arrived.

That changed on a ten-day meditation retreat in the California desert. It was day seven and participants were seventy minutes into a focus exercise, trying to pay attention to bodily sensations without passing judgment. But Siegel was overwhelmed by sensations. After a week of cross-legged meditation, his back ached, his neck throbbed, and his thighs were numb. "It was an all-consuming pain," he explains, "and all I was doing was judging."

And then he wasn't. Something inside him shifted. The part of his brain that had been judging suddenly turned off. "It felt like freedom," he explains. "If pure freedom feels like anything, that's what I was feeling. It was the most clear, present, and aware I had ever been. And if I could be in extreme pain and still remain peaceful and clear, then I thought maybe other people could do this, too. In that instant, everything I believed about human potential shifted."

It felt like a life-changing realization. But when Siegel got home from that retreat and returned to his normal routine, he couldn't integrate what he'd learned. It didn't matter how dedicated he was: the meditation practices he was trying to use were designed in a different time and for a different world. "In the world I lived in," he says, "I was surrounded by technology and information that seemed to be pulling me in a very different direction."

That's when Siegel realized that meditation was simply a tool meant to provoke a very specific reaction in the brain, but it wasn't the only tool available. In fact, considering all the recent advances in brain science and wearable sensors, meditation was pretty low-tech. So Siegel decided to build better tools, birthing the field that has come to be called "enlightenment engineering."

One of his early prototypes converted heart rate into audio tone. He was building on older research that showed prayer, yoga, and meditation could produce clear changes in heart rate. "I wore the thing for three days straight," recounts Siegel, "even when I slept. It was annoying. Beep, beep, boo, beep—and on and on. But, at the end of that period, just by working with this very thin slice of audio feedback, I learned to control my heartbeat. I could move it from forty beats per minute up to eighty and back again. It wasn't much more than a novelty toy, but it showed me what might be possible."

Since then, research has exploded. "Scientists all over the world are exploring contemplative practices," Siegel explains. "They're mapping a territory. And a whole slew of researchers, myself included, have begun using that map to create what you could call 'tech-assisted self-awareness devices,' or devices that can help us tune our internal environments."

In his work with heart rate variability, Siegel's found that by upgrading the tone to include a visual display, and adding in an EEG layer—so there's neurofeedback to go along with the biofeedback—he can get whole groups of people to synchronize their heart rates and brainwaves and drive them into group flow. His new challenge is to take this same technology and make an affordable version that's available to everyone.

He's also been moving beyond heart rate investigating ultrasound, transcranial magnetic stimulation and transcranial direct stimulation. These devices shoot pulses into the brain, and can turn on and off cortical regions with relative precision. "Right now," he says, "it's early days. So we have stuff that can make you feel like you've drunk a glass of wine. It's not quite what decades of meditation training can produce, but it's legitimate, reliable, and repeatable state change."

Since Siegel was living in Silicon Valley, he was obliged to form a company, Consciousness Hacking, around these ideas. He, alongside Nichol Bradford and Jeffery Martin, also cofounded the Transformative Technology Conference and started organizing consciousness-hacking meet-ups. In about nine months, with exactly zero spent on marketing, what began with a handful of people in one Northern California location has become a network of more than ten thousand people in twenty-three locations worldwide. In June 2015, his efforts scored him a feature in *The New Yorker.* Stanford University has taken notice, too—Siegel is now teaching courses on this emerging field to undergraduates.

"For the past three hundred years," Siegel explains, "there has been a split between science and religion. But now we have the ability to investigate this domain and innovate around spirituality. And whether you're judging by the growth of our meet-ups, the millions of dollars hitting this market, or the technology that's already available, lots of us are really interested in spiritual innovation.

And, if Siegel's predictions are correct, we've barely scratched the surface. "Consciousness-hacking technology is going to become as dynamic, available, and ubiquitous as cell phones. Imagine what happens if we can use personal technology to shift these experiences on demand, to support and catalyze the most important changes we can make at scale. More and more it's looking like we can retune the nervous system of the entire planet."

The Flow Dojo

Across the board, we're seeing an explosion in technologies that provide more people with more access to ecstasis than before. We now have sports equipment that gives mere mortals the chance to cheat death and chase flow, sound systems that entrance and entrain hundreds of thousands of people at once, immersive art that transforms waking reality into an interactive dream state, and biohacking tools that steer us towards tech-enabled transcendence. Each of these breakthroughs makes stepping outside of ourselves, easier, safer, and more scalable than ever.

Despite these developments, there's still a lot of untapped potential. Part of the trouble is that all of these disciplines—sport, music, art and biotech—have distinct subcultures and favored applications. While it's fairly common to see a couple of these combined—music and visuals are an obvious pairing, as are athletics and wearable sensors—putting them all together into one deliberately designed experience isn't typically done.

But it can be. As powerful as these advancements are independently, when blended together, their impact is amplified considerably. It's why, over the past several years, we've been collaborating with some of the top experience designers, biohackers, and performance specialists to help develop the Flow Dojo— a training and research center explicitly designed to merge these technologies in one place. Equal parts Cirque du Soleil, X Games, and hands-on science museum, it's a learning lab dedicated to mapping the core building blocks of optimum performance.

In the fall of 2015, we had the opportunity to bring a prototype of the Dojo to Google's Silicon Valley headquarters and engage in a joint-learning project. For six weeks, a handpicked team of engineers, developers, and managers committed to a flow training program, and then capped that off with two weeks in a beta version of the training center.

The premise was simple: if you train your body and brain, and manage your energy and attention, you'll be able to get into flow more frequently and perform better at work and at home. Each day, participants engaged in a range of activities, from sleep tracking, to diet and hydration, to functional movement (designed to undo the imbalances of deskbound lives), to brain entraining audio and respiration exercises. With just those basic practices, subjects reported a 35 to 80 percent increase in incidents of flow during their workdays. The bigger surprise for the engineers was that they also experienced more flow at home, where family dynamics were frequently less rational and predictable than the algorithms they played with at work.

Once that foundation was in place, we got to the interesting part—the Dojo itself. In our research for *The Rise of Superman*, we had interviewed more than two hundred professional and elite adventure athletes to figure out their secret to getting into flow so readily. Time after time, they told us it came down to two things: the right triggers and gravity.

And that's where the team began the design of the Flow Dojo. Was it possible to use technology to simulate those conditions in a safer and more accessible way? Could we take a page out of Alan Metni's iFly book and re-create the embodiment, and consequences that the world's best flow hackers relied on, to give regular folks a taste of the state? By combining Tony Andrew's sound design, Android Jones' digital imagery and Mikey Siegel's brain tech, could we construct a novel and interactive environment? If so, we could guide users into peak states, and researchers could capture invaluable data along the way. This would provide a unique opportunity to study the impact of wearables, experience design, and user biometrics, all in the same place. We could literally start reverse-engineering the genome of peak performance states.

So, we brought together a team of engineers to develop kinetic training gear that could deliver those experiences—think extreme playground equipment built for grown-ups. Giant looping swings that send you upside down and twenty feet off the ground, and pull more than three g's when you push through the arc's bottom. Momentum-powered gyroscopes and surf swings, complete with Doppler sound effects and LED cues, that let you flip, spin, and twist without risking a hospital visit.

The designers also integrated sensors and audio-visual feedback into the gear, so users get real-time data on physics (like g-force, RPM, and amplitude) and personalized biometrics (like EEG, HRV, and respiration). Taking that kind of data off smart watches and laptops—and away from the conscious mental processing of the prefrontal cortex—gets users out of themselves and into the zone with less distraction.

Even so, when Sergey Brin, one of Google's cofounders, stepped up to the looping swing, we were unsure how it was going to go. Brin is an action sports enthusiast, pursuing everything from BASE jumping to kitesurfing. At the TED conference a few years ago, he also topped the leaderboard on an EEG mindfulness training demo. So, while he already had some experience in both the physical and mental elements of this training, he had never put the two together.

First, we hooked him up to a heart rate variability monitor to establish a cardiac baseline. Then he strapped into the kiteboard bindings and began pumping the looping swing higher and higher. Most people encounter two limits on this device. The first is when they reach the highest point they ever attained on a playground swing, typically about 50 degrees. If they can push past that last known safe zone, the next limit is actually when they're straight upside down and needing to drive their weight forward (against all better judgment when you're twenty feet off the ground) to push the swing through a full revolution.

Brin pumped past both of those limits, looping the swing (only about 5 percent of subjects manage that feat on their first attempt), and then proceeded to stall the swing at its apex, and loop it blind and *backward*. When we compared his biofeedback data from his sessions to his baseline, he had lost coherence when he was initially exerting himself, but had regained an organized brain and heart rhythm once his body-knowledge figured out how to adapt and adjust. His only comment when he returned back to earth? "I want one in my backyard."

Brin's instinct to keep training these skills is supported by the data. Embodied cognition research shows that we become more flexible and resilient when we train our bodies and brains together, and in increasingly dynamic situations. It's why the SEALs say "you don't ever rise to the occasion, you sink to your level of training" and then proceed to overtrain for every scenario possible. It's a more advanced corollary to Amy Cuddy's power-posing advice: Once you get the basics down, start upping the ante. Try remaining centered under more challenging conditions (like managing heart and brain activity while swinging upside down). If we want to train for stability in all conditions, the science suggests, it's essential to practice with instability first.

Later in the visit, Google's other founder, Larry Page, gave one of Mikey Siegel's newest creations a try: a mix of 3D surround sound and visual feedback designed to prompt connection between people. Sitting in an enclosed dome, he and his wife put on small backpack subwoofers (so they literally felt the bass through their bodies, not their ears). They then watched two digital flowers blossom and contract on the screen surrounding them. But there was a trick to the setup—Larry was feeling his wife's heart beat and watching his flower pulse to her heart's rhythms, and she was watching and feeling his. By deliberately crossing the feedback loops, the installation creates technologically mediated empathy, no talking required. So absorbing was the experience that when the nighttime sprinklers came on and accidentally sprayed them, they just assumed it was part of the simulation.

While the field of immersive experience design and training is in its infancy, early results, like this project at Google, suggest that by combining all of the advancements in technology (movement, sound, light, and sensors) with an embodied hands-on training program, you can trigger a range of nonordinary states with far more precision and with much less risk. In the past, to get a glimpse of "no-self," it might have taken a high-risk wingsuit flight, a decade of monastic isolation, or a heroic (and possibly reckless) dosage of an unpredictable substance. Today, we can use innovations like the Flow Dojo to skillfully tweak and tune the knobs and levers of our bodies and brains and get similar breakthroughs with a fraction of the breakdowns.

"And really, that's the true power of technology and the four forces in general: they give more people more access to ecstasis in safer and more approachable ways. Without the shift in psychology, the notion of harnessing altered states toward practical ends would have seemed crazy. But we now know they can heal trauma, amplify creativity and accelerate personal development. Without advances in neurobiology, mystical experiences would have remained just that, mystifying. But we now know the precise adjustments to body and brain function that let us re-create them for ourselves. Without the progress in pharmacology, our exploration of nonordinary states would've remained constrained by geography, church and state. But we now know that a wide range of compounds disclose potentially revelatory information and insight. Without the developments in technology, too few were forced to risk too much to glimpse the value inherent in altered states. But we now know how to prime and prompt these experiences safely and at scale.

Driven by these changes, our understanding of ecstasis is now advancing at an exponential rate. Findings in one domain are informing and supporting developments in others. Research has been open-sourced, access has been democratized, and—as will become much clearer in Part III of this book—proof that these four forces are driving a revolution is everywhere you look.

PART THREE

The Road to Eleusis

The road of excess leads to the palace of wisdom.

—*William Blake*

CHAPTER EIGHT

CATCH A FIRE

The Sandbox of the Future

If gizmos and gadgets are your thing, then the annual Consumer Electronics Show in Las Vegas is your pilgrimage. If it's superheroes and graphic novels, then it's San Diego's Comic-Con. But if you're stalking ecstasis, if you want to see the four forces cranked up and deployed for full effect, then head out to that same desert festival Larry Page and Sergey Brin used to screen Eric Schmidt—Burning Man.

Every year, some three hours northeast of Reno, on a vast alkali flat known as the Black Rock Desert, you'll find all the major players from Part Two of this book. There's Tony Andrews in purple paisley, bumping bass out of a Funktion-One art car. Mikey Siegel's around, too, demonstrating neurofeedback to the dusty and the curious. Android Jones has erected a giant dome to display his trance-inducing visionary art. There are workshops by world-renowned sex therapists, lectures on neurotheology by Ivy League scientists, and every head-bending alphabetamine Sasha

Shulgin ever dreamed up. Off in the distance, you might even spot the Red Bull Air Force, dressed up in flaming costumes and wing-suiting into the city. Whatever else can be said about the event, Burning Man holds the undisputed title as the world's largest ecstatic trade show.

"Burning Man aggressively extends the tradition of hedonic ecstasy," writes Erik Davis. "[Wild] visuals, disorienting sonics, and a self-conscious excess of sensory stimulation . . . all help undermine stabilized frames. . . . [It's a] full-sensorium brain machine designed to bring us in tune with our mind's ongoing construction of real-time on the fly."

Michael Michaels, one of Burning Man's original founders (and known as "Danger Ranger" at the event), explains it this way: "At Burning Man, we've found a way to break out of the box that confines us. What we do, literally, is take people's reality and break it apart. Burning Man is a transformation engine—it has hardware and it has software, you can adjust it and tweak it. And we've done that. We take people out to this vast dry place, nowhere, very harsh conditions. It strips away their luggage, the things they've brought with them, of who they thought they were. And it puts them in a community setting where they have to connect with each other, in a place where anything is possible. In doing so, it breaks their old reality and helps them realize they can create their own." In other words, it's a transformation engine tailor-made to invoke the selflessness, timelessness, effortlessness, and richness of STER.

Increasingly, that transformation engine has been producing real change in the world. And that's the point of this chapter. If the past section examined the emergence of the four forces, this one asks the next obvious question: is the radical inspiration the forces provide leading to practical innovation? Earlier we explored studies that demonstrated nonordinary states can meaningfully boost creativity and problem solving under controlled conditions. Here, we want to step outside the lab and see if ecstasis is helping solve wicked problems "in the wild." And Burning Man is perhaps the best place to begin this inquiry.

The first thing to note is who's showing up at the festival. Unlike in the Woodstock era, attendees are no longer just countercultural bohemians who have "tuned in, turned on, and dropped out." For sure, there are still plenty of punk anarchists, industrial artists, and warehouse denizens for whom life in the mainstream is a sometimes awkward fit. But today, the ranks of Burners, as attendees call themselves, include members of a high-powered subculture, a tech-nomadic glitterati that have access to capital, markets, and global communication platforms.

When Tim Ferriss mentioned that nearly all of the billionaires he knows in Silicon Valley take psychedelics to help themselves solve complex problems, Burning Man is one of their preferred locations to step out and go big. "If you haven't been [to Burning Man], you just don't get Silicon Valley," serial entrepreneur and longtime attendee Elon Musk noted in *Re/Code*. "You could take the craziest L.A. party and multiply it by a thousand, and it doesn't even get fucking close."

Among certain circles, mention of "the playa" or "Black Rock City" gains you instant camaraderie with those who have shared that baptism by fire. Participation in successful Burning Man camps has morphed from countercultural street cred to career-building material. "So embedded, so accepted has Burning Man become in parts of tech culture," wrote journalist Vanessa Hua in *the San Francisco Chronicle*, "that the event alters work rhythms, shows up on resumes, is even a sanctioned form of professional development—all signs that the norm has adopted parts of the formerly deviant happening."

Over the last decade in particular, the festival has become a regular stop for those whose calendar might include Davos, TED, and a slew of other high-profile gatherings. In 2013, John Perry Barlow, a fellow at Harvard Law School and former lyricist for the Grateful Dead, casually tweeted from Burning Man: "Spent much of the afternoon in conversation with Larry Harvey, Mayor of Burning Man and Gen. Wesley Clark, who is here." At one of the more infamous parties on the planet, countercultural royalty are hobnobbing with a former Supreme Commander of NATO turned U.S. presidential candidate.

Three years later, the actual president, Barack Obama, joked about the event at the White House Correspondents Association Dinner, saying: "Just recently, a young person came up to me and said she was sick of politicians standing in the way of her dreams—as if we were actually going to let Malia go to Burning Man this year. Was not going to happen. Bernie [Sanders] might have let her go. Not us."

If the President of the United States is moved to comment on the event, and Elon Musk is claiming it's central to Silicon Valley culture, then perhaps there's more going on than just a weeklong party. And that's the second thing to explore in our assessment—why so many creative and talented people go so far out of their way to congregate there once a year. By simple elimination, it can't just be the sex, drugs, or music. Those indulgences, however tantalizing, are little more than commodities in any major city. There must be a pretty compelling *something* that inspires people to take a week out of their calendars to wander around an inhospitable salt flat in the middle of absolute nowhere.

Recent research conducted on Burning Man sheds some light on that "something." In 2015, a team of scientists led by Oxford neuropsychologist Molly Crockett joined forces with the Black Rock City Census to take a closer look at the festival's power. In their study, 75 percent of attendees reported having a transformative experience at the event, while 85 percent of those reported that the benefits persisted for weeks and months afterward. That's an incredibly high batting average: Three out of four people who attend the event are meaningfully changed by it.

And this doesn't just happen by accident. Wandering out into the middle of that intentional chaos at 2 A.M., surrounded by fire-spewing dinosaurs, giant neon-lit pirate ships, and the throbbing beats of galactic hip-hop, you're ripped away from all familiar reference points, totally unstuck in time, and well beyond normal awareness. The wildness of the event, the radical self-reliance it requires, the ability to create and inhabit larger-than-life alter egos, all combine to create a *temporary autonomous zone*—a place where people can step outside themselves and become, if only for a brief week, whatever they desire. It's the single greatest concentration of state-altering technology on the planet, designed by everyone together and no one in particular.

Which brings us to the final and most important category in our assessment: the astonishing amount of innovation this event consistently inspires. Attendees treat the playa as an oversized sandbox—a place where ideas can be dreamed up, tested out, and, as often as not, shared freely with everyone. "I like going to Burning Man," Google founder Larry Page said at the 2013 Google I/O conference. "[It's] an environment where people can try new things. I think as technologists we should have some safe places where we can try out things and figure out the effect on society, the effect on people, without having to deploy it to the whole world."

In 2007, Elon Musk did just that, debuting an early prototype of his Tesla electric roadster at the event. He also came up with the ideas for both his renewable energy company SolarCity and his superfast transit system Hyperloop while on the playa. And true to the Burning Man principle of gifting, he gave both away. SolarCity went to his cousins; Hyperloop, published online in a white paper, was an offering to the world at large (that has since inspired two different companies).

Zappos founder and CEO Tony Hsieh told *Playboy* that the experience of collective awareness, what he calls "the hive switch," is the reason he attends. That "feeling of unity with the other people in the space, unity with the music and with one another . . . that's why I go to Burning Man." For Hsieh, the festival has had such an impact that he's built its ideas into the corporate culture of Zappos, reorganizing the company to make flipping "the hive switch" as easy as possible. Similarly, he's also spearheaded the Downtown Project, an attempt to revitalize central Las Vegas with radical inclusion, interactive art, and other core elements of the festival.

While much has been made of the fact that Hsieh's efforts have faced setbacks and challenges, it would be surprising if things went differently. Hsieh has taken ideas pioneered at Burning Man and is attempting to reinvent the culture of a Fortune 500 company and to reinvigorate (to the tune of $350 million) a blighted urban core. That's structural change in the real world, with all the risks and complications it entails.

Musk's projects too, aren't without their complications. But reinventing transportation and pioneering a new energy grid (to say nothing of his efforts to colonize Mars) are wicked enough problems that they've stymied all prior efforts to solve them. What these examples make clear is that the perspective provided by nonordinary consciousness and culture offers a different path forward—a way to reconsider intractable challenges with fresh eyes.

All of these practical applications have, in turn, inspired the Burning Man organization itself. "A few years ago, we attended the event to speak at their annual TEDx series and then got invited to a small salon hosted by Danger Ranger. And it wasn't just Silicon Valley tech titans in attendance. Senior vice presidents from Goldman Sachs, heads of the largest creative ad agencies in the world, and leaders of the World Economic Forum, were all discreetly there, using fanciful assumed names, far from the flashbulbs and scrutiny of the media and the markets. Their goal was to forge a future based on the shared experience of communitas writ large: a permanent Burning Man community, a place where experiments with the four forces could be conducted year round.

As Burning Man cofounder Will Roger recently wrote: "I

would argue that the proposal is part of a large strain of utopian separatism that can be found in the modern-tech boom: Peter Thiel's Sea-steading efforts or Tony Hsieh's attempt to build a start-up city in Las Vegas. But a Burning Man permanent community would arguably be the most interesting and achievable manifestation of it."

In the summer of 2016, they achieved just that, closing on the purchase of Fly Ranch, a parcel of nearly four thousand acres a few miles north of the festival site, filled with geysers, hot springs, and wetlands. "This is all part of the evolution of Burning Man," the organization announced, "from an ephemeral experiment into a global cultural movement having an impact on social, economic and artistic norms and structures. Burning Man's culture is becoming more recognized and influential around the world."

When the Levee Breaks

One of the more interesting parts of this story isn't simply that Burners are trying to establish a homeland for ecstasis. It's that some of the hardest and grittiest lessons of building a city from scratch are showing up thousands of miles beyond the Nevada desert. So, if we want to continue our exploration of how nonordinary states are helping to solve wicked problems "in the wild," then we should head to some of the wildest places on earth—those ravaged by natural disasters and protracted war.

To better understand how a weeklong gathering could have such far-reaching impact, it's important to understand that in preparation for the event, all the central Burning Man organization does is survey the streets and put out port-a-potties. Everything else about the makeshift city—the camps, the giant art, the generators, the medical facilities, and the peacekeeping—is organized by volunteers. In coming together to create a city of seventy thousand, Burners are pioneering fundamentally different ways of organizing and mobilizing people in the face of some of the harshest conditions on the planet—and they're using the bonding power of communitas to do it.

One of the first times those skills truly got put to the test was in 2005. It was August 29, and over on the Gulf Coast Hurricane Katrina was less than an hour from making landfall. By the time the storm was over, it would spread $108 billion of damage from Florida to Texas, and earn the dark honor of being one of the five worst hurricanes in U.S. history.

Over in Nevada, meanwhile, it was sunny skies and light winds and Burning Man was in full swing. Camp PlayaGon (a combination of "*playa*" and "*Pentagon*") was bustling. A collection of high-ranking Pentagon officials, futurists, and hackers, PlayaGon had been charged with setting up and running the livestream broadcast and emergency Wi-Fi for the entire festival. But when news of Katrina reached them, they took a break from their duties to get a closer look.

"One of our guys took over a recon satellite," recounts Dr. Bruce Damer, a University of California biomedical engineer and NASA contractor. "Our Pentagon wireless satellite phone rang, the general on the other side was saying 'what's going on' and instructing our guy not to answer. We then had control of this thing and could watch Katrina come in." And never ones to miss an opportunity for a high-tech prank, the PlayaGon crew lit dozens of hydrazine flares (military-spec glow-sticks) around their camp and programmed the satellite to track the blaze from space, too.

But those real-time feeds of Katrina lashing the Gulf Coast had a sobering impact. The citizens of PlayaGon wanted to help. So did plenty of other Burners. After gathering more than forty thousand dollars in relief money from concerned attendees, an advance team left the festival, drove down to the Gulf Coast, and got to work.

With Doctors without Borders as their inspiration, they dubbed their nascent organization Burners without Borders. At the time, New Orleans was getting all the national attention, but these Burners decided to focus their efforts on coastal towns in Mississippi, which had been equally damaged but largely ignored.

The first thing they did was set up shop in a parking lot and build a much-needed distribution center for established charities like Oxfam and the Red Cross. Then, over the course of eight months, they donated more than $1 million in debris removal and reconstruction efforts. The organization did everything from restore a Vietnamese temple in Biloxi to raze and rebuild the entire town of Pearlington. As CNET noted: "This was no ragtag group of 10 to 20 hopeless do-gooders showing up without a plan. This was more than 150 people, toting heavy equipment, supplies of food and water, [and] years of experience surviving and thriving in harsh, off-the-grid environments."

Before leaving, they teamed up with local residents to build a giant sculpture out of flood debris and, true to form, turned it to ash in a giant, cathartic bonfire. "Our town was destroyed and we were abandoned by our government and our leadership," one Pearlington resident said, "but [Burners without Borders] came in and reminded us that even in all that devastation was the chance for art, for celebration and for community."

Since that time, Burners without Borders has become an international organization, active in disaster zones from Peru's 2007 earthquake to Japan's Fukushima disaster to New Jersey's Hurricane Sandy. And the relationships they've forged with locals in those areas have come full circle—with leaders from those communities coming to Burning Man in subsequent years to learn where all of that capability and enthusiasm comes from.

In an even less likely example of Burning Man's spreading impact, Dr. Dave Warner exported its core ethos to war-torn Afghanistan. In 2011, Warner, a data-visualization expert and the guy who had hacked that Katrina satellite, was in Jalalabad, just thirty miles from the caves of Tora Bora where Osama bin Laden had given U.S. forces the slip a decade earlier. A large man with long, graying hair and beard, Warner has a resume that reads:

"former U.S. Army drill instructor . . . PhD neuroscientist, technotopian idealist, dedicated Burner, dabbler in psychedelics, insatiable meddler and (weirdest of all) defense contractor."

Warner and a gang of MIT scientists, who called themselves the Synergy Strike Force, had posted up in Jalalabad to spread "the gospel of open information." Based on the Burning Man principle of radical inclusion, Warner insisted that all Synergy Strike Force projects remain unclassified and that the information be shared with everyone. "I'm dismantling the Death Star," he told a war reporter, "to build Solar Ovens for Ewoks."

So, Warner opened a "Burner bar," where he traded free drinks for terabytes of information. It was more of a Tiki hut, really—covered in bamboo, a simple cooler with some Heineken and the odd bottle of liquor displayed, but also a sign that read "We share information, communication (and beer)."

In their intelligence gathering, no detail was too small: reconstruction projects, troop movements, construction plans, hydrology surveys, health clinic locations, polling sites, names of local farmers, even crops those farmers were planting. Warner took all the information from his "Beer for Data" program (as it came to be known) and plugged it into a data-visualization tool he had created. The results outperformed every three letter agency you can think of, and—because Warner had refused a security clearance—he could "gift" anyone who asked with these results.

Lots of people asked. The Pentagon relied on his data, but so did the United Nations, Afghan officials, aid workers, and journalists. In one of the most chaotic environments in the world, gifting, transparency, and radical inclusion saved lives and dollars.

While Burners without Borders and Beer for Data mark two of the earliest examples of festival principles being exported into crisis zones, they're unlikely to be the last. "With so much experience in self-organizing their own municipal infrastructure in a hostile environment," former Apple executive Peter Hirshberg wrote in his book *From Bitcoin to Burning Man and Beyond*, "Burners are particularly skilled at functioning during chaotic crises when normal services—running water, electricity, communication channels and sanitation systems—are not available. Burners don't just survive in such an environment; they create culture, art and community there."

It's for this reason that Rosie von Lila, a former head of community affairs for the Burning Man organization, has been invited to the Pentagon three times and the United Nations twice to discuss infrastructure and disaster planning. "I've been amazed at the depth of interest," Von Lila says. "[T]raditional organizations are realizing the limitations of top-down mobilization and are seriously studying how bottoms-up community mobilization—the core lessons of the Burning Man community—can be deployed in critical environments."

Or, really, in any kind of environment. Burning Man "demonstration projects" can be seen everywhere from solar power installations on rural Indian reservations (Black Rock Solar) to experimental community spaces in blighted metropolitan areas (The Generator, in Reno, Nevada) to smartphone apps (including Firechat, which was designed as a peer-to-peer communication network at Burning Man, but then played a critical role in protest movements in Taiwan, Hong Kong, and Russia). And because Burners vigorously defend an open-source, noncommercial approach, their efforts are easy to share and hard to censor.

Burning Man "regional burns" now occur in nearly thirty countries, from Israel to South Africa to Japan, providing global access to the experience. It's been called a countercultural diaspora, but that might be too limiting. After all, what's countercultural about disaster relief, intelligence-gathering, and urban planning?

These projects all provided creative solutions to persistently wicked problems, ones that defied the best laid plans of the most powerful militaries, governments and aid agencies on the planet. Relying on the ingenuity, collaboration, and relentless hard work of a community forged by ecstasis, Burners are extending their impact well beyond the celebration that birthed them.

"Burning Man didn't invent the festival, the art car, or the Temporary Autonomous Zone any more than Apple invented the personal computer," continues Hirshberg. "But like that other venturesome innovator . . . Burning Man executed the concept beautifully, and through its work is having an outsized impact on our culture—and quite possibly on our future."

Disrupting the Brahmins

While Burning Man principles and skills are being deployed in some of the harshest conditions on earth by volunteers with limited budgets, a large number of the examples in this chapter have involved the creative class—that is, people with the resources, influence, and time needed for such pursuits. And, typically, that's the way these things have always worked.

At least as far back as the Eleusinian Mysteries, which counted notables such as Plato and Pythagoras among its members, ecstatic culture has often been spread by an educated elite. In Europe, we saw this with the Rabelaisians of the sixteenth century, and the Club de Hashish in the eighteenth century—both of whom explored altered states, open sexuality, and libertine philosophies in pursuit of inspiration. In the 1920's socialite Mabel Dodge Luhan's Taos home served as a mescaline-fueled salon for everyone from D. H. Lawrence to Georgia O'Keeffe and Carl Jung. In the 1960's Esalen's founders and faculty infused bohemia with academia, drawing heavily from the ranks of Stanford, Harvard, and the European intellectual community. Yet, even though these movements all began with a select few, they ended up having a disproportionate impact on philosophy, art, and culture.

In July 2013, we experienced a contemporary instance of this dynamic in the mountains of Utah, where a small but influential group of innovators are building community based on what they call "the power of shared peak experiences." Guided by Jon Batiste and his New Orleans marching band, we found ourselves in a large crowd of artists, activists and entrepreneurs, traipsing through an aspen forest. After about thirty minutes, the procession arrived at a wildflower-strewn meadow and the largest dinner table either of us had ever seen. It was a quarter-mile long—a single straight line stretching across the whole of the hillside—with white linen place settings for a thousand.

As we sat down, we noticed all the little details expressly designed to prompt awe and delight, like hand-crank radios playing jazz from a pirate AM station and stainless steel whiskey flasks with Walt Whitman poetry inscribed on their side. The hosts proceeded to serve everyone by hand, laying out an inventive multicourse meal. Then, timed perfectly with the rising full moon to the east and the blazing sun setting in the west, everyone raised a glass to the palpable sense of community present that night.

After dinner, the entire forest transformed into an LED wonderland. Soundstages were playing everything from pulsing electronica to spoken word poetry. Butterfly and glo-worm art cars buzzed soundlessly up and down the dirt paths. And in the distance, scattered across the hillside, were the domes, tents, and pavilions that housed everyone gathered to connect and collaborate over this midsummer weekend.

As much as the fingerprints of Burning Man were everywhere at this event—the remote setting, the glamour camping, the art, performance and whimsy—there were two crucial differences: not only did it take place at nine thousand feet above sea level, but it wasn't going to disappear at the end of a week. This summertime gathering was a coming-out party. Just seven weeks earlier, the hosts, Summit Series, had bought the entire mountain.

"We wanted a permanent home," explains Summit cofounder Jeff Rosenthal. "We wanted to build a town dedicated . . . to what altered states really can provide: creativity, collaboration, innovation, entrepreneurship and community. And because our community shared that vision, we were able to crowdsource $40 million and buy a ski area (Powder Mountain) that sits on a mountain range the size of Manhattan." So while folks at Burning Man are just starting to build themselves a homeland, Summit has already taken that step.

Already, there are more than five hundred home sites on the land, with people like Richard Branson, Kobe Bryant, GE's CMO Beth Comstock, and Nike president Trevor Edwards already committed to the project. And instead of the typical McMansion resort plan, they are actively fostering community by prohibiting oversize statement homes and concentrating development into tightly clustered neighborhoods. Everything is being built to platinum-level LEED environmental standards. It's the world's first ecstatically inspired eco-town, though it didn't start out that way.

Summit began in 2008, when five entrepreneurs in their early twenties came together to solve a common problem. They didn't know any *really* successful entrepreneurs, and had no one to ask for advice. So the quintet came up with a creative solution: cold-call business leaders and ask them to go skiing.

Nineteen people showed up, including Zappos's Tony Hsieh and Facebook cofounder Dustin Moskovitz. "We learned that when you take a bunch of really bright, diverse people," explains Rosenthal, "and let them share a dynamic immersive experience, you get powerful results. Lifelong friendships were formed. It removed the tedious, transactional nature of networking. I guess you could say that one of the things we discovered on that trip was that altered states accelerate business."

This one-off experiment soon became the Summit Series, a string of "nonconferences" built upon those original insights. The series, which has been called "TED crossed with Burning Man" or "the hipper Davos," has struck a chord. Summit's first event was a ski trip for nineteen people, their second event was a trip to Mexico for sixty, and their third event was at the White House.

After hearing about the organization, President Obama invited them to bring thirty-five young leaders to dinner for a discussion about Millennial culture and the future of innovation. And what they shared at the White House, and have been implementing ever since, is a vision of social entrepreneurship that values both purpose and profits.

Today, when Summit hosts a weekend, you'll find little-known entrepreneurs, activists, and artists mingling with the likes of Questlove, Eric Schmidt, and Martha Stewart. This kind of cross-pollination has produced some interesting collaborations. Summit mobilized a team of marine biologists, adventurers, and philanthropists to organize a protect-our-oceans trip through the Caribbean and raised more than $2 million to establish a nature preserve. They've supported Pencils of Promise, a nonprofit dedicated to global K–12 education that's built almost four hundred schools, and helped launch Falling Whistles, a global network of over 120,000 members dedicated to eradicating child soldiers in the Congo.

But it's not just nonprofit work they're interested in. They've also begun a venture capital fund that has helped seed dozens of start-ups, including the buy-one/give-one shoe company TOMS, the eyeglass upstart Warby Parker, and the ride-sharing giant Uber. By using non-ordinary states to promote community, they're reimagining the staid world of professional networking, philanthropy, and venture capital.

And Summit isn't the only organization leveraging those lessons to accelerate change. MaiTai Global, started in 2006 by venture capitalist Bill Tai and kitesurfing legend Susi Mai, uses action sports (mostly surfing and kitesurfing) as a stimulant for group flow and entrepreneurship. Theirs is a potent partnership. Bill Tai sits on the boards of a half-dozen of Silicon Valley's best-known companies, while Susi Mai is the only woman to be awarded lifetime status as a Red Bull athlete (one of the ultimate honors in action sports).

MaiTai hosts multiday gatherings that blend kiteboarding sessions, off-the-record conversations with founders, start-up pitch marathons, and a transformational festival atmosphere. "We curate our experiences very strategically," explains Susi Mai. "We find the right mix of really interesting people and subject them to powerful state-changing experiences that accelerate social bonding. It's the same formula used at Burning Man and at Summit."

Mai also points out that there's a lot of crossover between these cultures. "Very early on, we got a lot of support from the Burning Man community. Burners instinctively understood what we were trying to do so they just started showing up. And it's a very participatory community, so when Burners do show up, they build stuff, they start organizing, they get everyone involved."

Very involved. To date, dozens of companies have leveraged the talent and contacts in the community to raise venture funding and find key partners, most done with handshake deals on the beach, at the end of a great kiting session, when, as Tai says, "everyone's feeling the stoke." They've also come to see action sports as a filter for start-up success. "We've noticed that learning to kitesurf has a lot of parallels with the challenges of entrepreneurship," Mai explains. "If someone has the grit and presence of mind required by the sport, then it shows a lot about their character and how they show up more broadly in life." And the parallels aren't just conceptual. Over the years, MaiTai members have founded and led companies with an aggregate market value of more than $20 billion, making them one of the more influential (and athletic) groups of entrepreneurs in the world.

Building on these insights, MaiTai recently created the Extreme Technology Challenge (abbreviated to XTC, naturally). Rather than having promising start-ups pound the pavement in the hunt for funding, the Challenge gathers them together at the

Consumer Electronics Show in Las Vegas for a pitch-fest meant to invert this process. "Instead of droves of founders showing up for a series of thirty-minute meetings with venture capitalists," explains Tai, "and those funders then having to 'backdoor vet' prospective leaders, we spend real time together in an environment where a person's true character is revealed."

Finalists receive an invite to Richard Branson's Caribbean hideaway, Necker Island, where, between kitesurfing sessions, they get to pitch Branson himself. On Necker, as at Burning Man and Summit's Powder Mountain, everything is deliberately designed to create communitas. We experienced this firsthand when we were invited to the island to talk about flow and entrepreneurship. From the cliffside zip lines that carry you to breakfast, to the stunning Balinese-inspired architecture, to a broad menu of action sports, everything is built to trigger that state of effortless focus. "When I do [reach flow]," Branson told us over smoothies on his back porch one morning, "I get an extra two hours of great work done, and the other twelve are really, really productive—so trying to get that balance in life is really important; not saying that one shouldn't party hard as well."

Branson and Mai Tai are taking this same approach to host the Carbon Warroom, a transnational organization dedicated to energy sustainability in the Caribbean, and the Blockchain Summit, an international consortium exploring socially beneficial applications for alternative currencies. By bringing the passionate and talented together to play and work, they're charting a course toward a more innovative and sustainable future.

Case in point: a cedar hot tub perched on the crow's nest of the main house of Necker Island. Guests gather there to soak, talk, and gaze at the stars, and it was at one of those late-night gatherings that Branson birthed his most ambitious company yet. "That's where I had the idea for Virgin Galactic," he reflects. "NASA hadn't yet created a spaceship I could fly on and, if I waited too long, I wouldn't be around. So, I thought, let's build our own. I mean, who in their right mind wouldn't look up at those stars and not dream of going there?"

And getting innovators into their "right minds" is what Summit, MaiTai, and Branson have done so well. By realizing that non-ordinary states are more than just a recreational diversion and can, in fact, heighten trust, amplify cooperation, and accelerate breakthroughs, a new generation of entrepreneurs, philanthropists and activists is fundamentally disrupting business as usual.

High Times on Main Street

If the only evidence of the four forces showing up in the world was to be found at exotic gatherings for a fortunate few, their impact would remain severely constrained: "trickle-down ecstasis." But what is emerging is more varied than that. The ripple effects of these innovators' companies and projects are even beginning to show up on Main Street itself—and they're following a predictable pattern of dissemination.

In his seminal book *Crossing the Chasm,* Geoffrey Moore outlined exactly how new ideas gain traction. At first, when breakthroughs happen, only those people willing to tolerate the risk and uncertainty of a novel technology get on board, a trade they'll make for the benefits of being "early adopters." Then there's a gap, what Moore called "the chasm," that any idea has to cross to attract a growing audience. It's attracting that "early majority" on the far side of the chasm that he feels is the true mark of disruptive innovation.

Up until now, we've focused primarily on the pioneers and early adopters—those most visibly driving the evolution of the four forces. Here we want to take a moment to catalog signs of broader applications, focusing on the places where the chasm has been crossed and a critical early majority are starting to incorporate state-changing tools and techniques into their everyday lives.

Take the first force, psychology. Thanks to the work of Martin

Seligman and others, a new generation of positive psychologists is repackaging meditation, stripping out its spiritual connotations, and providing evidenced-based validation for its benefits. This new version, known as mindfulness-based stress reduction, is gaining traction in places that would never have embraced earlier variants. Eighteen million Americans now have a regular practice, and, by the end of 2017, 44 percent of all U.S. companies will offer mindfulness training to employees. Since rolling out their program, Aetna estimates that it's saved $2,000 per employee in health-care costs, and gained $3,000 per employee in productivity. This quantifiable return on investment helps explain why the meditation and mindfulness industry grew to nearly $1 billion in 2015. What had been the domain of seekers and swamis is now a staple of HR.

And the impact of positive psychology is spreading well beyond the workplace. At Harvard, professor Tal Ben Shahar's course on happiness is the most popular in the university's history, while mainstream books on the science of well-being consistently top the bestseller lists. And this focus on optimal living isn't just bettering our mood; it's advancing our growth. One of Bob Kegan's graduate students recently determined that by college, many Millennials have reached stages of adult development (with all their associated increases in capacity) that took their parents until middle age to attain.

We're seeing similar progress in neurobiology. Legitimized by discoveries in embodied cognition, contemplative physical practices like yoga, tai chi, and qigong have become the most popular indoor activities in the United States. Consider yoga. This five-thousand-year-old tradition was a countercultural pastime until the 1990s. But once researchers began finding the practice did everything from improve cognitive function to decrease blood pressure, the general public started to cross the chasm. As of 2015, some 36 million Americans have a regular practice. An activity that changes our state of mind by changing the shape of our bodies has become more popular, in terms of participation, than football.

On the higher-tech end of the spectrum, state-changing treatments like transcranial magnetic stimulation are now outperforming antidepressants, and many Silicon Valley executives are going off-label, using the technology to "defrag" their mental hard drives and boost performance. Companies like Dave Asprey's Bulletproof Executive are helping people biohack their daily lives with everything from smart sensors to nootropics (brain stimulating supplements). This market is expanding so rapidly that Bulletproof has grown into a nine-figure enterprise in less than four years and hundreds of other companies are flooding into the market.

In pharmacology, we're increasingly accepting of substances that shift our consciousness. Marijuana, once called a "demon weed," has become the fastest-growing industry in America. The whole of the cannabis economy (including legal and medical) is now worth roughly $6.2 billion, and slated to rise to $22 billion by 2020. As of late 2016, twenty-eight states have legalized medical marijuana, and eight of them—Colorado, Washington, Oregon, Nevada, California, Massachusetts, and Alaska, and the District of Columbia—have legalized recreational use as well. Over the next five years, researchers believe another fourteen states will follow suit. As Peter Reuter, a University of Maryland drug policy professor, recently told CNN, "I'm surprised by the long-term increase in support for marijuana legalization. It's unprecedented. It doesn't look like a blip."

And cannabis is merely the most obvious sign of this change. Whether we're examining psychedelics like LSD or empathogens like MDMA, mind-altering drugs are more popular than at any other time in history. Thirty-two million Americans use psychedelics on a regular basis (that's nearly one in ten) and report considered reasons for doing so. According to a 2013 study published in a journal of the National Institutes of Health, the most common motivations are to "enhance mystical experiences, introspection and curiosity." Transcendence, not decadence, appears to be driving use forward.

Technology has seen similar developments. A few decades ago, brain imaging devices were multimillion-dollar machines available in only a handful of university labs. Today they've become as ubiquitous and accessible as the smartphones in our pockets. With a handful of plug-and-play sensors, we can now measure hormones, heart rates, brainwaves, and respiration and get much clearer pictures of our real-time health.

In the summer of 2016, for example, Jay Blahnik, the lead designer of the Apple Watch, gave us an early look at their product road map. Over the next few years the watch will connect these sensors to become a platform for open-source research into everything from obesity to peak performance. In one twenty-four-hour beta test, more than thirty thousand people volunteered to contribute their personal data to Alzheimer's research, making it four times the size of the next-largest study overnight.

And Apple is only part of a larger trend. Between 2000 and 2009, companies filed fewer than four hundred patents for neurotech. That number doubled in 2010, and doubled again in 2016. With the data these devices are providing, we can shortcut our way not only to better health, but to deeper self-awareness, taking weeks and months to train what used to take yogis and monks decades to master.

While these are all examples of the four forces reaching deeper into the mainstream majority, they may already seem unremarkable. There's a reason for this. Ray Kurzweil, the director of engineering at Google, once pointed out that it's hard for nonscientists to track progress in artificial intelligence because, when it shows up in the real world, "it looks like nothing fancier than a talking ATM."

It's true for ecstasis as well. Soccer moms with Kundalini yoga practices; business men microdosing psychedelics; tech nerds tracking biometrics, *The Simpsons* going to Burning Man—these developments might seem pedestrian. But they are the "talking ATM's" of altered states. They are proof that the chasm has been crossed, that the once cutting-edge is now integrated into the everyday.

Nothing New Under the Sun

Under the hot August sun, in the western wilds of America, tens of thousands of misfits gather to worship and celebrate. These seekers are there because they reject the stuffiness of their parents' religion but are equally uninspired by the godlessness of their transient society. They crave direct mystical experience, and they have come together to find it.

They stay up all night dancing, playing music, getting intoxicated, and crowding together to see the headlining performers. Standing on giant scaffolds, these artists whip the crowds into a collective trance. When the mood takes them—and it often does—attendees have sex under the open sky.

Afterward, when they go back to their regular lives, they transform the world as they know it. On fire with their recent initiation, they challenge existing social, political and spiritual conventions. So noticeable are their efforts that the towns and cities where they congregate are called "burned-over" districts.

There was a time and a place when this all happened, but it wasn't the present day and it wasn't the Black Rock Desert. The date was 1801; the place was Cane Ridge, Kentucky. The occasion was the Second Great Awakening, one of the largest spiritual revivals in American history.

The ink was barely dry on the Constitution, and the western boundary of the United States reached only as far as the Appalachians, but already the foundation of a vibrant American ecstatic tradition was being laid. Those gathered in Cane Ridge were part of the largest revival of that era. More than twenty thousand settlers camped out, listening to itinerant preachers who stood on elevated platforms to speak to the frenzied crowds of the coming Rapture. In between sermons, people hung out by their tents and lean-tos, with fiddles and banjos, playing the Scots-Irish tunes that would later become bluegrass music. And despite the ostensibly pious intent of the gathering, there was plenty of drinking and fornicating. Even back then, the "Holy Ghost feeling" was tough to keep under wraps.

These revivals offered connection and community in a world that felt fragmented and hectic. Over the next half century, an entire generation of the young and passionate joined in. The Second Great Awakening gave birth to social justice movements ranging from temperance and women's rights to abolition. It infused American politics with an activist conscience for years to come. Even Joseph Smith's hilltop epiphany took place in one of those "burned-over" districts.

So as we consider the emergence of the four forces and where they lead, it can be helpful to realize that the revolution we're experiencing today might be more the norm than the exception. American spirituality has always favored the direct over the inferred, the immediate over the gradual. It has always spilled over from the pews and pulpits into the towns and countrysides.

In this context, we could consider this current moment as a Great Awakening in its own right. Only this time, the mythical has been replaced by the empirical. From the Nevada desert and the disaster zones of the Gulf Coast and Afghanistan to the mountains of Utah and the sidewalks of Main Street, people are coming together to see for themselves. And what they're discovering is that there's more capacity, resilience, innovation, and creativity in all of us collectively than in any of us alone. That's as significant today as it was over two hundred years ago. So even if there really is "nothing new under the sun," each time it rises, it's still a sight to see.

CHAPTER NINE

BURNING DOWN THE HOUSE

Even though a critical mass of the population may be crossing the chasm, and incorporating the benefits of nonordinary states into their lives and work, that doesn't mean this revolution won't cause problems. Historically, every time ecstasis has shown up, it's led to upheaval and misuse. That's because, while the insights provided by the four forces may give us a better way to stabilize these experiences and lessen that risk, there will always be those who try to bend them to other ends.

Back in the section on Pipers, Cults and Commies we touched on these dangers, examining the pitfalls that nonordinary states can pose for individuals and groups—namely, the dual issues of coercion and persuasion. Here we're going to expand that thread by focusing on two of the institutions with the most vested interest in coercion and persuasion today: the military and marketers.

We'll start with militarization, reviewing more than half a century of government attempts to weaponize consciousness. Then we'll move into commercialization, where the power of ecstasis is being used to open our wallets. This latter category is a more recent development, to be sure, but one with a high potential for abuse. In both cases, we'll see how the application of nonordinary states, as with other powerful technologies, has both ethical and political ramifications.

The Atomic Donkey

It was 1953 and the Pentagon had a problem. Colonel Frank Schwable, a U.S. Marine Corps pilot, had been shot down over North Korea, appeared on Chinese radio, and confessed that he'd been ordered to deploy biological weapons. The event was a PR nightmare. If the Pentagon let Schwable's story stand, they'd be caught in violation of the Geneva Convention; if they discredited his account, they'd be undermining a decorated officer and prisoner of war.

So the Secretary of Defense demanded, as Annie Jacobsen recounts in her recent book, *The Pentagon's Brain,* "an all-out campaign to smear the Koreans [with] a new form of war crime, and a new form of refinement in atrocity techniques, namely mind murder, or 'menticide.'" If Schwable had been the victim of communist mind murder, then his testimony could be invalidated and his patriotism upheld—a tidy solution to a messy problem.

Menticide, most in the Pentagon agreed, was a clunky word. But the CIA had been discreetly testing a more compelling tagline in *New York Times* op-eds: "brainwashing." This one stuck. Brainwashing neatly encapsulated one of the deepest fears of the Cold War era—the idea that your very individuality, your own free will, could be hijacked by a totalitarian state.

The CIA sowed the specter of brainwashing so successfully in the minds of the American public and within its own operational culture that it came to be considered one of the primary threats of the Cold War. So even though they'd dreamed up this bogeyman themselves, perfecting mind-control devices and drugs to combat it became a top-secret, top priority.

Not long after Schwable's radio announcement, the Department of Defense got wind that a brilliant young University of Pennsylvania neuroscientist might have discovered the very technology they'd been seeking. Representatives from nearly every government agency—the CIA, NSA, FBI, Army, Navy, Air Force, and the State Department—all beat a path to Dr. John Lilly's door.

Lilly had solved the two biggest technical problems of mechanically inducing ecstasis on demand. The first was that inserting electrodes through the skull and into the brain invariably caused too much damage. The second was that pulsing unidirectional current across nerve endings tended to irreparably cook the circuitry.

But Lilly had developed tiny stainless-steel sleeves you could tap into a subject's skull and then slip gossamer electrodes through, with virtually no swelling or lasting harm. He'd also built a machine that sent bidirectional electrical pulses through the brain that stimulated neurons without knocking them out of balance. The procedure itself was virtually painless—nothing more than pinpricks as the sleeve guides went in. The electrodes could be inserted to any depth in the brain, from the cortex down to the amygdala. And the guides could remain embedded and undetected for months or even years.

In primates, Lilly had discovered that the pleasure system—what could be called the brain's basic ecstatic circuitry—correlated directly with the sexual arousal network. Male monkeys trained to use his device for self-stimulation would choose to orgasm nonstop for sixteen hours, followed by eight hours of deep sleep, after which they would get right back to it. Pleasure, Lilly had discovered, was an endlessly motivating and potentially all-consuming pursuit (at least in males).

For this reason, when the director of the National Institute of Mental Health (NIMH) told him to brief the Pentagon on his work, Lilly expressed concern. "Anybody with the proper apparatus can carry this out on a human being covertly," he recounted in his autobiography, *The Scientist*. "If this technique got into the hands of a secret agency they would have total control over a human being and be able to change his beliefs extremely quickly, leaving little evidence of what they had done."

To guard against this, Lilly detailed a series of nonnegotiable conditions under which he would be willing to discuss his findings. Nothing he said could ever be classified and everything shared would remain experimentally repeatable by him or his colleagues. Long before Linus Torvalds gave away the source code to Linux, or Sasha Shulgin published his chemical cookbook, or Elon Musk shared all of Tesla's car and battery patents—long before there was even a term for it—Lilly took a stand for open-sourcing ecstasis.

What he hadn't counted on was how relentless the military could be. Not long after that initial presentation, Lilly was contacted again, this time by an unnamed representative of the Sandia Corporation (a Lockheed Martin subsidiary and longtime defense contractor). He wanted "to learn the technique of inserting the sleeve guides into the heads of large animals." Again, Lilly insisted on keeping the work declassified, but agreed to let the man come and film his latest experiments.

A few years later, *Harper's Magazine* wrote an in-depth piece on Sandia detailing their "super mule" project—a donkey/horse hybrid equipped with electrode implants and a solar compass. The mule carried its load, quite likely a suitcase nuke, in an exact straight line, regardless of terrain. If it veered, it was punished with pain. If it tracked, it was rewarded with pleasure. As he read the piece, Lilly was shocked to recognize a photo of the man who had filmed his experiment. Sandia had managed to take mechanically induced ecstasis and harness it to wage nuclear war.

Devastated, Lilly realized that before he could complete his research, government agencies were going to co-opt it. He disavowed experimenting on animal or human test subjects, concluding that self-experimentation was the only ethical way to explore the boundaries of the mind. He left the NIMH and ceased all research with "neuro-physiological aids." Yet, despite his abandoning his position and his funding, and risking his reputation and ultimately his life, Lilly's work would prove endlessly fascinating to the military and intelligence communities for decades to come.

He Who Controls the Switch

In 2010, Tim Wu, a professor at Columbia Law School, discovered that information technologies, ranging from the telegraph to radio, movies, and ultimately, the internet, tend to behave in similar ways—starting out utopian and democratic and ending up centralized and hegemonic. In his book *The Master Switch,* Wu calls this "the Cycle," a recurring battle between access and control that shows up whenever these breakthroughs emerge. "History shows a typical progression of information technologies," he explains, "from somebody's hobby to somebody's industry; from jury-rigged contraption to slick production marvel; from a freely accessible channel to one strictly controlled by a single corporation or cartel—from [an] open to closed system."

When radio operators began stringing up towers in the early 1920's, for example, it was so people could talk to each other and share ideas over an open broadcast medium. "All these disconnected communities and houses will be united through radio as they were never united by the telegraph and telephone," wrote *Scientific American.* But that's not what ended up happening.

By the mid 1920's, AT&T and RCA teamed up to create the National Broadcasting Corporation (NBC), controlling access to bandwidth and creating a massive multinational company that persists to this day. By the 2000's, another juggernaut, Clear Channel Communications, controlled market share and playlists in more than thirty countries. This was unification, for certain, but not of the democratizing variety imagined by the early pioneers.

Because of the inevitability of the Cycle, Wu believes there's no question more important than who owns the platform—the means by which people access and share information. It's what prompted him to coin the term "*net neutrality*" back in 2003 and spawn an ongoing conversation about the balance of civic and corporate power online. It's also where he got the title of his 2010 book. "Before any question of free speech," he writes, "comes the question of 'who controls the *master switch*?'"

While information technologies started out concrete and physical—ranchers putting up telegraph wire to connect their farms to town, and radio stations building giant AM antennas—they're getting increasingly virtual: the ones and zeroes of the internet and the infinite complexities of Google's search algorithms. And with the four forces, information technology is moving from the virtual to the *perceptual.*

Ecstatic technology isn't limited to silicon chips and display screens. As John Lilly's early research established, it's the knowledge of how to tweak the knobs and levers in our brain. When we get it right, it produces those invaluable sensations of selflessness, timelessness, effortlessness, and richness. And that final step—the richness? That's the information that we can't normally access. As W. B. Yeats put it, "The world is full of magic things patiently waiting for our senses to grow sharper."

Once information technology become perceptual—as in the case of nonordinary states—the Cycle becomes even more powerful. Our mind becomes the platform. The tug-of-war between access and control becomes a battle for cognitive liberty. And while nation states have consistently sought to regulate external chemicals that shape consciousness, what happens when they attempt to regulate internal neurochemistry?

If that sounds far-fetched, consider that elite athletes already submit "biological passports" to the World Anti-Doping Agency to confirm their unique baselines for hormones, blood profiles, and neurochemicals. If they fluctuate from that baseline without official permission, they are penalized and even brought up on criminal charges. Much in the same way that regimes used to declare certain books subversive, it's not too much of a stretch to imagine a government declaring certain brain chemistry subversive. A telltale combination of neurotransmitters coursing through your bloodstream could be enough to get put on a watch list, or worse.

So while it's tempting to herald the four forces as a development that is going to unlock ecstasis for the masses, we'd be naïve to think that a persistent historical pattern—the battle for control of the Master Switch—won't apply this time around.

Spooks to Kooks

The struggle over Lilly's brain stimulation device was an early example of the Cycle in action—of whether an ecstatic technology could remain freely accessible, or would end up centrally controlled. Since then, that struggle has evolved into a decades-long game of cat-and-mouse between the "spooks" of the intelligence community and the "kooks" of the counterculture. Scientists like Lilly repeatedly pioneered new techniques to alter consciousness just in time to have the government attempt to weaponize them. Or the spooks worked on some new top-secret application, only to have it leak out and get repurposed by the kooks. And while some of the stories we'll cover in this section may sound so outlandish they stretch credulity, they consistently underscore Wu's thesis, the high-stakes game of who controls the Master Switch.

It turns out that more than a few of those Pentagon officials who came knocking on Lilly's door were funded by the CIA. They were part of MK-ULTRA, arguably the largest and most notorious brainwashing project in U.S. history. Some eighty institutions, including universities, colleges, hospitals, prisons, and pharmaceutical companies, took part. Their goal was to find chemicals that could control and confuse enemy combatants, civilian populations, and heads of state—including one *Spy vs. Spy* plan to slip Fidel Castro an LSD-soaked cigar.

"Within the CIA itself, [agents] were taking LSD regularly, tripping at the office, at Agency parties, measuring their mental equilibrium against those of their colleagues," Jay Stevens recounts in *Storming Heaven*. "Turn your back in the morning and some wiseacre would slip a few micrograms into your coffee. It was a game played with the most exalted of weapons, the mind, and sometimes embarrassing things happened. Case-hardened spooks would break down crying or go all gooey about the 'brotherhood of man.'"

In addition to these frat-boy antics, the program engaged in more serious lapses in judgment. They repeatedly dosed mental patients and prompted one of their own, a chemist at Fort Detrick's Biological Weapons Center, to jump or get thrown (the evidence is conflicting) out a thirteenth-story New York City hotel window. And, in the annals of unintended consequences, MK-ULTRA gets a notable mention for accidentally unleashing a leviathan: the psychedelic revolution of the 1960s.

Almost exactly twenty-five hundred years after Alcibiades's first stole *kykeon,* a young student named Ken Kesey poached some too—only this time it was from the CIA. Like Alcibiades, Kesey was disarmingly persuasive and controversial, wangling his way to a tuition-free spot in a graduate writing seminar at Stanford and enduring a criminal trial and exile of his own. Just as Socrates had doubted whether Alcibiades was a worthy pupil, Wallace Stegner, the literary lion who headed the writing department at Stanford, didn't think much of Kesey, either. Stegner dismissed him as "a sort of highly talented illiterate" and "a threat to civilization and intellectualism and sobriety." Which as it turns out, wasn't far off.

As background research for his novel *One Flew Over the Cuckoo's Nest,* which was set in a mental institution, Kesey had been volunteering at a U.S. Veterans Administration hospital (which, unbeknownst to the young author and many of the administering doctors, was part of MK-ULTRA). To earn a little extra money, a friend of his had turned him onto the $75 per session experiments the docs were running there on "psychomimetic" drugs—meaning chemicals like LSD that mimicked the mental breakdown of psychosis. The scientists "didn't have the guts to do it themselves," Kesey later told *Stanford Alumni* magazine, "so they hired students. When we came back out [of the sessions], they took one look at us and said, 'Whatever they do, don't let them go back in that room!'"

Over on Perry Lane, the bohemian cottage enclave where he lived, Kesey and his growing band of pranksters took things out of the lab and into the field. "Volunteer Kesey gave himself over to science at the Menlo Park vets hospital," Tom Wolfe recounts in *The Electric Kool-Aid Acid Test,* "and somehow drugs were getting up and walking out of there and over to Perry Lane."

"Half the time," Wolfe continues, "Perry Lane would be like some kind of college fraternity row with everybody out on a nice autumn Saturday afternoon on the grass . . . playing touch football . . . an hour later Kesey and his circle would be hooking down something that in the entire world only they and a few other avant-garde neuropharmacological researchers even knew about."

What happened next became the well-documented subject of counterculture lore. Kesey moved the experiment into the hills above Palo Alto, Hunter S. Thompson, the Hells Angels, and Neal Cassady (from Kerouac's *On the Road* fame) all showed up, as did a strange little band called the Grateful Dead, led by a chinless but oddly magnetic guitarist named Jerry Garcia. Armed with gallons of day-glo paint, strobe lights, and the prototypical art car, a tricked out 1939 International Harvester bus named "Further," Kesey and his Merry Pranksters birthed West Coast psychedelic culture. Control of the Master Switch had been wrestled away from the spooks, and neither Silicon Valley nor the wider world would ever be the same.

Over the next decade, Eastern mysticism, liberated sexuality, and "following your bliss" mounted a direct challenge to the traditional values of mainstream America. "But, while the kooks were enthusiastically sharing esoteric techniques with a broader audience, the spooks never lost interest in the movement they had accidentally birthed. By the mid-1970's, Watergate had broken, Saigon had fallen, and a demoralized Defense Department was in serious need of inspiration. "A round of post-Vietnam soul-searching," *Fortune* reported, "culminated in the establishment of Task Force Delta, a cadre of army officers whose mission was to scan for new ideas." No one was better at scanning for those ideas than Jim Channon, a lieutenant colonel in the Army and veteran of two tours of duty in Vietnam. "I just made it my weekend duty to get around all of these places, like Esalen, make friends and find out what this esoteric technology really was."

By the time he'd finished his hot tubs and crystals junket, Channon had, for all intents and purposes, gone native. He penned *The First Earth Battalion Operations Manual,* making the case that deliberately cultivating nonordinary states, including the ability to experience universal love, to perceive auras, to have out of body experiences, to see into the future, and, perhaps most memorably, "to encounter the enemy with sparkly eyes"—could transform the military.

And as far out (and dated) as this sounds, Channon's manifesto took on fabled status among progressive thinkers in the military. In "Beam Me Up Spock: The New Mental Battlefield," a 1980 article for the staid journal, *Military Review,* Lieutenant Colonel John Alexander argued that "a new battlefield dimension that may defy our general perceived concepts of time and space looms on the horizon. Clearly, psychotronic (mind/matter) weapons already exist; only their capabilities are in doubt." Even the U.S. Army's famous "Be All You Can Be" slogan sprang from Task Force Delta's mission to unlock human potential.

A couple of years later, the Pentagon commissioned the Trojan Warrior Project, an intensive six-month training in mind-body-spirit practice for Green Berets. The program included meditating with a Tibetan lama, neuro and biofeedback sessions in a cutting-edge computer lab, praying with a Benedictine monk, and training in aikido, a Japanese martial art dedicated to universal peace. It was a frontal assault on the neurophysiology of ecstasis (and the direct progenitor of the SEALs' Mind Gym). For their coat of arms, they combined ancient and pop mythologies: a wooden horse sat above two crossed light sabers. Their motto? *Vi Cit Tecum*—"May the Force Be with You."

While this progressive era produced some undeniable "white hat" dividends—ranging from mindfulness and stress reduction programs for the general enlisted to martial arts training in the Marine Corps—there were also some "black hat" applications. In his manual, Channon had lobbied for the calming, soothing, and inspiring capacities of music, hoping that bass, not bombs, would prevail on the battlefield of the future. Almost as an afterthought, he'd added that, if all else failed, "unpleasant, discordant sounds could be used to disorient enemy combatants."

But that afterthought got noticed. In May 2003, *Newsweek* ran a short blurb "PSYOPS: Cruel and Unusual," revealing that U.S. military detention units were using a combination of bright light, disorienting sounds, and other consciousness-shifting tactics to break Iraqi prisoners. "Trust me, it works," says one U.S. operative. "In training, they forced me to listen to the Barney 'I Love You' song for 45 minutes. I never want to go through that again."

That was the sound bite that launched a thousand clips. But rather than acknowledging the military's ethically questionable interrogation tactics, the news cycle spun happily on, with TV hosts inserting a "Barney is torture for us too" gag right in between footage of pandas at the zoo and the local weather. What began as an attempt to infuse the military with the idealism of the human potential movement had devolved into a tool for psychological warfare—and the Cycle churned on.

And it's still churning today. Consider the government's clandestine role at Burning Man. On the surface, the festival—a one-week gathering on an utterly forgettable patch of U.S. Bureau of Land Management desert—is not what you'd consider a "high-value target." But for the short few days of its existence, the event holds the dubious distinction of being one of the most surveilled cities in the country. Despite experiencing less violent crime than most midsize suburbs, it draws over a dozen separate state and federal agencies equipped with millions of dollars of high-tech spy gear, infrared goggles, tactical vehicles, and undercover agents.

In heavily redacted documents recently released through the Freedom of Information Act, it turns out that the FBI has conducted a multiyear intelligence program at Burning Man. The official reason was to scout for domestic terrorists and track potential threats from Islamic extremists. More likely, the FBI was taking a page out of their old COINTELPRO playbook, the one used in the 1960's to infiltrate and destabilize the Black Panthers, Students for a Democratic Society, and the American Indian Movement. If that were the case, then one would expect increased surveillance of the event, heightened policing, insertion of agents provocateurs, and aggressive prosecution of nonviolent crimes. And while it's hard to tell if it's an anomaly or the beginning of a trend, in 2015, plainclothes and undercover agents spiked, and arrests at the festival were up 600 percent.

It seems safe to say that the intelligence community knows something big is happening out in the desert, they just can't quite figure out what. That's because, other than the obvious external cues—the fiery explosions, wild costumes, and all-night dance parties—what's really going on is happening in people's minds. To the rank-and-file law enforcement monitoring, "the festival, it must seem like a rowdier Mardi Gras, or a Times Square New Year's with fewer drunks and more hugging." But not so for the top brass. In some instances, as we saw when Camp PlayaGon commandeered a spy satellite, and the Supreme Commander of NATO attended the event, they're in on it.

And this repeated pattern of the "spooks lying down with the kooks," from hippie float tanks at the SEALs' Mind Gym, to Kesey's misadventures at the V.A. hospital, to Lieutenant Colonel Channon hottubbing at Esalen, to the Pentagon at Burning Man, clearly highlights the back and forth contest for control of the Master Switch. More critically, it illustrates one of the central challenges of ecstasis: how to ensure that powerful techniques for altering consciousness don't get used for the wrong reasons.

To note that "a tool is morally neutral" is a standby of college philosophy papers, but in the case of ecstatic technologies, it's unsettlingly true. As we saw in earlier chapters, fully expressed ecstasis tends to promote empathy, compassion, and well-being. But at 80 percent expression? What then?

Even this brief survey of the past half century shows that ecstasis can easily be bent to darker ends. The selflessness that is the hallmark of a nonordinary state is only a hop, skip, and a jump from the brainwashing the Pentagon so desperately sought in the 1950's. Timelessness, devoid of reference points, can feel a lot like paranoid schizophrenia and has been a linchpin of solitary confinement for centuries. The euphoric neurochemistry of effortlessness, as John Lilly realized, can create dependency on whoever can administer that next hit of bliss. Information richness can be mined as a truth serum, as the MK-ULTRA docs attempted, or amped up to overwhelm the unwilling, as the military guards orchestrated in Iraq.

In the same way that it takes a far less developed society to detonate a nuclear bomb than to invent one, the power of ecstasis constantly tempts those who would have no idea how to replicate it on their own. But once they see it in action, once they can map the fundamental logic, it doesn't take much to turn it to ends that would mortify its original creators.

Soma, Delicious Soma!

If the prospect of the military-industrial complex hijacking ecstasis to pursue its own agenda is sobering, an equally likely outcome is that we end up seduced by our own desires. In fact, control not through coercion—as totalitarian states have done—but through persuasion is an even more likely prospect.

In 2007, a collection of the world's biggest brands—Apple, Coca-Cola, American Express, Nike, Samsung, Sony, and Ford—put up $7 million to fund a study into the neuroscience of buying behavior. They wanted to know if there were more effective ways to sell their products and joined forces to underwrite the largest neuromarketing study ever conducted—an attempt to replace misleading focus groups with straight-ahead brain scans.

Marketing researcher and consultant Martin Lindstrom teamed up with British neuroscientist Gemma Calvert for the project. Over the course of three years, they used both fMRI and EEG to scan the brains of more than two thousand people as they made a variety of buying decisions. The researchers discovered that product placement in TV shows and movies rarely works, that warning labels of cigarettes actually prime the urge to smoke more, and that—most surprisingly—shopping and spirituality seem to rely on similar neuronal circuitry.

When deeply religious subjects view sacred iconography or reflect on their notion of God, brain scans reveal hyperactivity in the caudate nucleus, a part of the pleasure system that correlates with feelings of joy, love, and serenity. But Lindstrom and Calvert found that this same brain region lights up when subjects view images associated with strong brands like Ferrari or Apple. "Bottom line," Calvert reported, "there was no discernible way to tell the difference between the ways subjects' brains reacted to powerful brands and the way they reacted to religious icons and figures. . . . Clearly, our emotional engagement with powerful brands . . . shares strong parallels with our feelings about religion."

Lindstrom's high-profile advocacy of the neuromarketing revolution put him on *Time*'s list of the "100 Most Influential People." But it triggered a backlash. Critics rightly pointed out that just because spiritual symbols and corporate logos activate similar brain regions, doesn't make shopping a religious experience. While Lindstrom may have exaggerated the capabilities of neuromarketing in 2007 (he is, after all, a marketer), by the next decade the idea of tweaking the knobs and levers of the brain for purely commercial ends had become much more of a reality.

In 2013, for example, we were asked to keynote the annual

meeting of the Advertising Research Foundation. A global consortium of just about every major brand you can think of— from Coca-Cola, Wal-Mart, and Procter & Gamble to creative agencies like J. Walter Thompson, Ogilvy & Mather, and Omnicom, to tech giants like Facebook, Google, and Twitter—the foundation wanted to learn about the use of flow in advertising. Could this state of consciousness play a role in prompting buying behavior? Could the mechanics of ecstasis be used to drive market share?

To understand this possibility, it's helpful to understand a few of the developments that have led to today's marketplace. At the tail end of the twentieth century, we started moving from the selling of ideas, the so-called information economy, toward the selling of feelings, or what author Alvin Toffler called the "experience economy." This is why retail shops started to look like theme parks. Why, instead of stocking ammo on their shelves like Wal-Mart, the outdoor retailer Cabela's turns their stores into a hunter's paradise of big-game mounts, faux mountainsides, and giant aquariums. It's how Starbuck's can charge four dollars for a fifty-cent cup of coffee: because they're providing that cozy "third place" between work and home.

But we were at the Advertising Research Foundation to discuss the next step: the move from an experience economy to what author Joe Pine calls the "transformation economy." In this marketplace, what we're being sold is who we might become—or as, Pine explains: "In the transformation economy, the customer IS the product!"

On the surface, the idea that we would favor products that could help us become who we want to be doesn't sound bad. Take the fitness industry. In the experience economy, one of the undisputed leaders is Equinox Gyms, which blends state-of-the art equipment, boutique lobbies, and eucalyptus steam baths to create a luxury workout. You may or may not get as lean as those models in the black-and-white photo spread, but you'll certainly feel like a million bucks while you're there.

In the transformational economy, CrossFit charges almost as much but offers none of those perks. Instead, what you get is the promise that after three months of sweating in their stripped-down boxes (as CrossFitters call their workout spaces), you'll become a radically different person. You'll look different, for certain, but because of their emphasis on embracing challenge and pushing boundaries, you'll stand a chance of acting and thinking differently as well. That's a positive "transformation" that many are willing to suffer and pay a premium for.

Yet, it doesn't take much to bend this desire for personal change in more commercial directions. Consider a recent Jeep campaign, where they built mud bogs at county fairs. With thumping music and flashing lights amplifying the joyride, Jeep let fairgoers hop into one of their stationary rigs, floor the motors, spin the tires, and send dirt flying. The novelty of the experience; the rapid shift in sensations; the lights, music, and cheering crowd, was all more than enough to trigger the brain's pleasure machinery and get red-blooded twenty-somethings fixating over no-money-down leasing options for weeks to come.

That Jeep campaign worked so well because it effectively created a state of peak arousal for its participants and then sold them on an imagined transformation of their lives (starting with the purchase of a 4x4). Under those amped-up conditions, salience—that is, the attention paid to incoming stimuli—increases. But, with the prefrontal cortex down-regulated, most impulse control mechanisms go offline too. For people who aren't used to this combination, the results can be expensive.

The video game industry may have gone further down this path than anyone. "Games are a multi-billion dollar industry that employ the best neuroscientists and behavior psychologists to make them as addicting as possible," Nicholas Kardaras, one of the country's top addiction specialists, recently explained to *Vice*. "The developers strap beta-testing teens with galvanic skin responses, EKG, and blood pressure gauges. If the game doesn't spike their blood pressure to 180 over 140, they go back and tweak the game to make it have more of an adrenaline-rush effect. . . .

Video games raise dopamine to the same degree that sex does, and almost as much as cocaine does. So this combo of adrenaline and dopamine are a potent one-two punch with regards to addiction."

Armed with knowledge of our deepest longings, and an understanding of exactly how to prime them, large corporations are at a distinct advantage in the influence game. In the same way that Google tailors searches based on our past histories and targeted ads follow us around the internet until we buy, we are entering an era where our cravings for transcendence can be used to co-opt our decision making.

Once you understand what Lindstrom calls "buyology," you can imprint unsuspecting consumers with all the pleasure-producing neurochemistry you can coax out of them. And as with the intelligence community's efforts, ecstasis at 100 percent is transformational, but ecstasis at 80 percent is, well, pretty much whatever you want it to be.

With the advancement of the four forces, finding ways to shape decisions we're not even aware we're making has become increasingly straightforward. Less than a year after our presentation to the Advertising Research Foundation, DARPA ran an experiment demonstrating just how simple this really was.

In their study, a trained storyteller told an audience wired to EEG sensors and heart rate monitors a heart-wrenching tale of childhood bullying. Then she asked for donations to an organization working to end this behavior. Simply by reviewing the biometrics, DARPA scientists were able to predict with 70 percent accuracy who was most deeply moved by the story and who would choose to give money to the cause. Physiological data alone was enough to predict future spending.

They also discovered how to prompt that impulse. For the pitch to be most effective—that is, to earn the most money—it had to be highly engaging and display significant contrast between positive and negative story elements. Since the speaker was wearing a discreet earpiece while onstage, the researchers could use biofeedback to provide instant feedback, telling her to change the story on the fly, increasing tension, deepening empathy, and constantly priming the audience to alter their behavior.

While this study focused on a relatively benign example of persuasion, the very fact that DARPA was the one funding it should give us pause. Imagine newscasters or politicians wielding similar technology, able to pluck heartstrings, stoke outrage, inspire hope, and even trigger communitas, just by reading and tuning our neurobiology. If "focus-group politics" leaves us with a bad taste, how will "biofeedback politics" go down?

Kevin Kelly, futurist and the cofounder of *Wired* magazine, has a few ideas. In a 2016 article on virtual reality darlings of the moment, Oculus and Magic Leap, Kelly examines VR's potential as a technology of surveillance and control. "It's very easy to imagine a company that succeeds in dominating the VR universe quickly stockpiling intimate data on not just what you and three billion other people 'favorite' but . . . a thousand other details. To do that in real life would be expensive and intrusive. To do that in VR will be invisible and cheap."

Soon VR systems are going to track everything from eye gaze to vocal tone to—as DARPA-style biometrics get further integrated—neurochemistry, hormones, brainwaves, and cardiac coherence. "This comprehensive tracking of your behavior inside these worlds," continues Kelly, "could be used to sell you things, to redirect your attention, to compile a history of your interests, to persuade you subliminally, to quantify your actions for self-improvement . . . and so on. If a smartphone is a surveillance device we voluntarily carry in our pocket, then VR will be a total surveillance state we voluntarily enter."

So imagine the kind of immersive visionary experience that Android Jones is creating, one already designed to prompt state change, then add in this kind of biofeedback loop. In exchange for the thrill of getting higher, we'll willingly give up intimate details about ourselves. It'll be the new cost of getting our minds blown.

Aldous Huxley's dystopian novel, *Brave New World*, gave us a look at exactly how this could happen. Set in the year 2054, Huxley described a hypercommercialized world in which people were conditioned with brainwashing, sexual diversions (in the forms of group "orgy-porgies"), and soma—a psychedelic antidepressant offering "all the advantages of Christianity and alcohol; none of their defects."

"In [George Orwell's] *1984* . . . people are controlled by inflicting pain," wrote NYU professor Neil Postman. "In *Brave New World*, they are controlled by inflicting pleasure. In short, Orwell feared that our fears will ruin us. Huxley feared that our desire will ruin us." And while the possibility of a nation deliberately invading our minds to shape and control behavior may feel like a relic of Cold War paranoia, the prospect of multinational corporations deliberately tweaking our subconscious desires to sell us more stuff is already here.

Ecstasy Wants to Be Free

So if these two dynamics—commercialization and militarization—are powerful enough to co-opt our deepest drives, what chance do we really have of maintaining our independence? To be sure, it's asymmetrical warfare. Compared to each of us finding our way one step at a time, governments and corporations have a much larger stake in and budget for controlling ecstasis. Playing by those old rules, we don't stand a chance.

In *The Master Switch*, Tim Wu acknowledges as much, describing the struggle over any information technology as an inevitable tug-of-war between nation-states and corporations—and that *either* of them, left unchecked, creates imbalances. States can overreach. Companies can monopolize. Instead, Wu calls for constraining "all power that derives from the control of information." "If we believe in liberty," he writes, "it must be freedom from both private *and* public coercion."

It's for this reason that so many of the Prometheans we've met in this book have taken a stand for open sourcing. When the government came knocking, John Lilly demanded his ideas remain declassified. When Sasha Shulgin got that first hint of a DEA crackdown, he published all his pharmacological recipes. It's there in the democratizing effects of Mikey Siegel's consciousness-hacking meet-ups, it's why OneTaste has built an Orgasmic Meditation app downloadable anywhere in the world, it's what fuels the volunteers of the Burning Man diaspora. Open-sourcing ecstasis remains one of the best counterbalances to private and public coercion.

And once we do take those freely shared ideas and use them to unlock nonordinary states for ourselves, what do we find? A self-authenticating experience of selflessness, timelessness, effortlessness, and richness. In short, all the ingredients required for a rational mysticism. It cuts out the middlemen, and remains rooted in the certainty of the lived experience. This ability to continually update and advance our own understanding, ahead of anyone else's attempts to constrain or repurpose them may be the key to breaking the stalemate.

Wu agrees. "The Cycle is powered by disruptive innovations that . . . bankrupt the dominant powers, and change the world. Such innovations are exceedingly rare, but they are what make the Cycle go." An open-source approach to non-ordinary states makes Wu's "disruptive innovations" a little less rare, and the ability to share and distribute them less susceptible to co-optation. And while the four forces don't guarantee a bloodless revolution, they do ensure that more people get to decide for themselves.

And that's the ultimate paradox of these states: all that liberation comes with an unavoidable dose of responsibility. While these states provide access to heightened performance and perspective, the upsides come at a cost. Between our own wayward tendencies and the dangers of militarization and commercialization, it's easier than ever to fall asleep at the Switch.

CHAPTER TEN

HEDONIC ENGINEERING

So if the responsibility to democratize ecstasis falls squarely on us, we need to remember that we're no longer protected by the pale. We're out here on our own. When it comes to exploring consciousness, as Sasha Shulgin used to say, "there are no casual experiments."

That's the goal of this final chapter, to provide a set of guidelines for training nonordinary states, and share a few key ideas from the research in this domain. We'll identify known dangers, places where the overzealous typically blow themselves up, and suggest a handful of solutions to the more common problems people encounter. Think of it as a user manual for ecstasis.

"Known Issues" of STER

At the end of the chapter on neurobiology, we introduced a potential upgrade for making sense of ecstasis: repurposing our egos from our operating system (OS) to a user interface (UI). Making this switch can help us unburden our psychology and manage the intensity of a wider range of states without overclocking our processors. But, when it comes to exploring those states, we still have to contend with a whole set of "known issues." Call them the downsides of STER. Exposure to the selflessness, timelessness, effortlessness, and richness of an ecstasis can go wrong, and wrong in predictable ways. For each of these experiences, there is a corresponding danger that, if we know about it ahead of time, we have a chance to avoid.

Selflessness: It's Not About You

You could call it a messiah complex and you wouldn't be wrong. Certainly, the messiahs come in droves. But so do Virgin Marys, King Davids, and at least one Samson—who proved his virility by smashing through a wall of the psychiatric hospital in Old Jerusalem. There are thousands of cases on record, pilgrims who visit the city and get kicked out of reality by the weight of all that holy history. Rather than deciding, "Wow, I just had a mystical experience where I *felt* like Jesus Christ!" they conclude, "Wow, I *am* Jesus Christ. Clear the decks, people, I've got things to do!"

First identified back in the 1930s, "Jerusalem Syndrome" is a temporary fit of madness brought on by a visit to one of the world's most sacred sites. It's an overdose of spiritual awe, where historical significance and religious potency team up to overwhelm the unprepared. Occasionally, it afflicts people with preexisting mental conditions; mostly it hits people with devout religious beliefs. Every now and again, it fells average tourists.

Psychologists call this reaction "extreme ego inflation." Often, the experience of selflessness is so new and compelling that it feels like no one else has ever felt this way before—that it's evidence of some kind of sacred anointment. When triggered by an awe-inspiring encounter with the Wailing Wall, the result is Jerusalem

Syndrome. But the same thing can happen with any ecstatic experience. It's why Burning Man advises people to not make any life-changing decisions for at least a month following the event, and why online psychedelic message boards like Erowid are filled with advice like "Don't believe everything you think."

In nonordinary states, dopamine often skyrockets, while activity in the prefrontal cortex plummets. Suddenly we're finding connections between ideas that we've never even thought of before. Some of those connections are legitimate insights; others are flights of fancy. In 2009, Swiss neurologist Peter Brugger discovered that people with more dopamine in their systems are more likely to believe in secret conspiracies and alien abductions. They're suffering from *apophenia,* "the tendency to be overwhelmed by meaningful coincidence," and detecting patterns where others see none.

When the prefrontal cortex shuts down, impulse control, long-term planning, and critical reasoning faculties go offline, too. We lose our checks and balances. Combine that with excessive dopamine telling us that the connections we're making are radically important and must be immediately acted upon—that *we're* radically important and must be listened to—and it's not hard to imagine how this goes wrong.

So no matter what comes up, no matter how fantastical your experience, it helps to remember: *It's not about you.* Take an encounter with selflessness for all the possibility it suggests, but fold those lessons back into your everyday roles and responsibilities. As Buddhist teacher and author Jack Kornfield reminds us, "after the ecstasy, the laundry."

Timelessness: It's Not About Now

In 1806, General Zebulon Pike set out to survey the Rocky Mountains. When he crossed into Colorado and first sighted the fourteen-thousand-foot peak that would later bear his name, he wrote in his journal that he expected to camp on its summit in two days, three days at most. Three *weeks* later, Pike and a small band of his men gave up climbing in defeat. Pike, himself, failed to climb Pike's Peak. And for an understandable reason: utterly unaccustomed to the thin air and high elevation, he'd badly misjudged the distance and the scale of the terrain he was exploring.

A similar issue arises when we encounter the timelessness of nonordinary states. Our ability to accurately estimate how close things are to happening, both in our life and in the world at large, can get seriously skewed. Under normal conditions, with an active prefrontal cortex constantly scanning scenarios in the past and the future, we spend very little time living completely in the present. So when a nonordinary state plunges us into the immediacy of the deep now, it brings an added sense of gravitas to the moment.

Popular religious movements, from the Seventh-Day Adventists in the 1840's to Mayan calendar adherents in 2012, have all bet (and lost), on putting a doomsday pin in the calendar. Contemporary psychonauts have even coined a term for this persistent distortion: *eschatothesia*—the perception of the Eschaton, or the end of the world. "It is not necessarily the absolute 'end of times,'" the Hyperspace Lexicon clarifies, "but can be a feeling of some huge event in the near future we are approaching, the end of an aeon, a marker in time after which nothing will be the same."

And it's not just would-be prophets who suffer from this distortion. Anyone who experiences the clarity and immediacy of ecstasis and tries to bring those insights back to reality has to account for the time lag. A surfer who in a flow state drops into a wave and strings together a series of moves he's never pulled off before may need months of hard training to be able to reproduce them in a contest. An entrepreneur who glimpses a brilliant business model while dancing at a festival may need years to build the company that actually delivers on it. A musician who hears a fully formed symphony in her head during a meditation retreat could take the rest of her life to become skilled enough to actually play it.

Which is fine if we anticipate it, but demoralizing if we don't.

"Most people overestimate what they can do in one year," Bill Gates once said, "and underestimate what they can do in ten." In bringing back ecstatic insights, it's critical that we calibrate the difference between the reach-out-and-touch-it immediacy of the "deep now" with the frustratingly incremental unfolding of the day-to-day. As Zebulon Pike learned the hard way, at high elevations, objects in the mirror are sometimes *much* farther away than they appear. Remember: *It's not about Now.*

Effortlessness: Don't Be a Bliss Junkie

With effortlessness, we see a different downside: the combination of all those reward neurochemicals and overflowing inspiration can be intoxicating. Once people taste the fleeting effortlessness of ecstasis, some decide that's how life is always supposed to be: a state of perpetual ease. They become bliss junkies, state chasers, refusing to do anything unless they can, to borrow a phrase, "go with the flow."

Think of yourself as a colander—a bowl filled with holes. When you experience a peak state, it's like turning on the kitchen faucet and flooding that colander with water. If there's enough volume, the colander fills up despite the leaks. As long as water keeps flooding in, you will, for a moment, experience what it's like to be a cup. You'll feel whole; if you're really inspired, holy.

Then the faucet turns off, the peak experience ends, and all that water leaks back out. In a matter of moments, you'll settle back to where you started. The information recedes. The inspiration that was so easy to grasp moments ago slips away. And now you've got a decision to make. Do you engage the dull and repetitive work of plugging your leaks or do you go hunting for the next ecstatic faucet to tap?

The notion that hard work and persistence in the face of struggle might have a role in all of this often gets lost. In 2014, Ryan Holiday released a bestselling book on exactly this subject, *The*

Obstacle Is the Way. It offered an update to the Roman Stoic Marcus Aurelius's claim that "the impediment to action advances action. What stands in the way becomes the way." And this is certainly true of the ecstatic way. All that "effortless effort" takes a lot of work.

So do the hard thing and the rest becomes much easier: Enjoy the state, but be sure to do the work. And no matter how tempting it is: *Don't become a Bliss Junkie.*

Richness: Don't Dive Too Deep

On November 17, 2013, Nick Mevoli bet on himself one last time, and lost. This was at Dean's Blue Hole, a nearly seven hundred foot deep patch of ocean in the Bahamas. Mevoli was a rising star in the free-diving community, where divers swim as deep as they can with only a lungful of air. He had shot to prominence in the prior couple of years, winning contests, setting records and finding solace in his life under the water. "Water is acceptance of the unknown, of demons, of emotions, of letting go and allowing yourself to flow freely with it" he had written in his blog. "Come to the water willing to be consumed by it, but also have confidence that your ability will bring you back." That day, after barely resurfacing from a seventy-two meter warm up dive into the Blue Hole, Mevoli went into cardiac arrest and died. This time, he wasn't able to bring himself back.

When asked to comment on the accident, Natalia Molchanova, regarded by many as the greatest freehold breath diver in the world, said, "the biggest problem with freedivers . . . [is] now they go too deep too fast." Less than two years later, off the coast of Spain, Molchanova took a quick recreational dive of her own. She deliberately ran though her usual set of breathing exercises, attached a light weight to her belt to help her descend, and swam downward, alone. It was supposed to be a head-clearing reset. But, Molchanova didn't come back either.

And that's the problem that free diving shares with many other state-shifting techniques: return too soon, and you'll always wonder if you could have gone deeper. Go too far, and you might not make it back. "When we go down," Molchanova had told an interviewer not long before her death, "we understand we are whole. We are one with the world." That feeling is called "the rapture of the deep," a euphoric high produced by alterations in the lungs' gaseous chemistry, and it's responsible for one in ten of all dive fatalities.

For those exploring nonordinary states, there's a similar danger. You can stay down too long, amazed at what you're discovering. You also can become enraptured by the deep. And if Mevoli and Molchanova are the cautionary tales of free diving, the poster child for ecstasis gone too far, is someone whose story we've come to know well—Dr. John Lilly. After abandoning his experiments at the National Institute of Mental Health, Lilly went deeper and deeper into his psychedelically fueled float tank research. By precisely balancing the effects of ketamine and sensory deprivation, he was able to blast off into the furthest reaches of inner space. So he could maintain contact with the *Matrix*-like reality he encountered there, Lilly started injecting himself with ketamine on an hourly basis, often for weeks at a time.

On one occasion the tank was too warm, but when he tried to get up and adjust the temperature, he fell back in and, immobilized by the ketamine, drowned. His wife, Toni, came into the lab to find her husband floating facedown and blue. She revived him using the CPR technique she had learned while reading the *National Enquirer* only two days earlier.

That wasn't enough to slow Lilly—the rapture of the deep kept calling. While ketamine is so benign it's routinely used as an anesthetic for children and pregnant women, it has an underground reputation as "the heroin of psychedelics." As Lilly discovered, it's the utterly novel information it provides, not its chemistry, that's so addicting.

Not long after that near-fatal drowning, he had another brush with death. In the hospital, Lilly had a textbook NDE and, as he reported afterward, was visited by the same entities he'd been encountering in the float tank. They presented him with a choice: leave with them for good, or return to his body, heal, and focus on more worldly pursuits. Finally, Lilly got the message. He abandoned his psychedelic research and retired with his wife to Hawaii, where he lived to the ripe age of eighty-four.

There may not be another researcher who has dived as deeply into the mysteries of consciousness as John Lilly. He may just have been lucky, or, as he believed, had some help from other realms. But undoubtedly his rigorous training as a scientist, his insistence on preserving a critical and objective stance in the face of even the most outrageous experiences, saved his mind, and possibly his life.

After all this, Lilly came to one overarching conclusion: "What one believes to be true is true or becomes true, within certain limits to be found experientially and experimentally. These limits are further beliefs to be transcended. In the province of the mind, there are no limits."

If he was right, and there really are no limits to consciousness, then the point is not to keep going until we find it all, but to come back before we've lost it all. Because it really doesn't matter what we find down there, out there, or up there, if we're unable to bring it back to solid ground. So take it all in, but hold it loosely. And most critically, *Don't Dive Too Deep.*

This leaves us with four rules of thumb to carry into our exploration of these states. *It's not about you* and *it's not about now* help us balance ego inflation and time distortion. While *don't become a bliss junky* and *don't dive too deep* ensure that we don't get seduced by the sensations and information that arise in altered states.

Are these the only "known issues?" Not even close. Check out any of the world's contemplative traditions or the *Diagnostic and Statistical Manual of Mental Disorders* and you'll find dozens more. But these four cautions tie directly to the four core qualities of altered states—STER. They're nonnegotiable. If you put it in the ditch with one of these, you've got no one to blame but yourself. "No sympathy for the devil," Hunter S. Thompson once wrote. "Buy the ticket, take the ride . . . and if it occasionally gets a little heavier than what you had in mind, well . . . maybe chalk it off to forced consciousness expansion."

The Ecstasis Equation

In skiing, the North Face of the Aiguille-du-Midi, a twelve-thousand-foot peak in Chamonix, France, is one of the ultimate challenges. "If the North Face . . . isn't extreme skiing, then nothing is," Hans Ludwig wrote in *Powder* magazine. "Every one of its routes features thousands of feet of complex high-angle terrain with constant death exposure, mandatory rappels, and a snowpack that sticks tenuously to the hanging shield of glacial ice crowning the peak." Despite centuries of ski mountaineering in the region, no one completed a descent until 1994. Then, in 2001, Kristen Ulmer decided to give it a try.

Known as one of the best multisport athletes in the world, Ulmer was an extraordinarily talented big mountain skier, ski mountaineer, rock climber, ice climber, and paraglider. With a level of dominance not often seen in any sport, she was named "Best Female Extreme Skier in the World"—twelve years in a row. Of course, she wanted to become one of the first women to ski the North Face.

Two nights before Ulmer's descent, four feet of snow fell. The next morning she got a late start. As the sun climbed, the snow started to soften and slide. "We got pinned on a 70 degree face, standing on a one-inch band of ice, while seven-story avalanches poured over our backpacks—for three hours. When it was over, they sent a crew in by helicopter to retrieve our bodies. They were pretty shocked to find us alive. I was pretty shocked to be alive."

This would have been a wake-up call for many, but Ulmer couldn't live without the rush. "Whenever I felt that level of fear, that's when I felt the most alive. I was totally in the zone. And I was addicted to that feeling, I couldn't quit."

So she didn't. She went skiing the next day, and again almost died. "That was followed by three more near-death experiences," she explains. "I had five in five months. It was pretty clear the universe was screaming at me to stop doing what I was doing—and, I thought, being as nice about it as possible. But being a pro athlete was my salvation, my identity and my career. I had put decades of work into it. Quitting just seemed ridiculous."

Then, in the middle of her crisis, Ulmer went to Burning Man for the first time. "I was blown away. I got the exact same feeling I got from sports. I got it from the interactive art. I got it from the group flow. And that was it, I quit my career cold turkey and walked away from everything." Or, as *Ski Journal* once explained: "In the 1990s, Kristen Ulmer blew up the freeski scene as the first true female extreme skier. . . . then, she disappeared."

Ulmer had realized that ecstasis can be accessed through a host of different pursuits. And that realization, as she says, "saved my life." Since making this discovery, she has designed and built some of Burning Man's most famous art cars (including the one Tony Hsieh moved to his Downtown Project in Las Vegas). She also trained as a meditation and performance coach, and now leads Zen ski camps. She hasn't had to risk her life for a thrill in years.

"I still measure the quality of my life by the number of times I get into the zone," explains Ulmer. "If I spend two weeks at Burning Man and only get that experience a handful of times, then I feel cheated. It wasn't worth it. But now I can try different things. That's the real change. Now I know I have options, that there are actual comparisons to make."

Knowing there are options matters to plenty of people besides Ulmer. SEALs speak quietly of this same dilemma, of how hard it is to flip the switch when they come off deployment. We hear similar accounts in action sports meccas like Squaw Valley and Jackson Hole. Consistently we're approached by athletes and their loved ones looking for ways to take fewer risks. Yet, like Ulmer in Chamonix, they haven't figured out any other way to feel quite as alive.

Invariably, in those same conversations, someone always asks, "but what's the *best* way to get into the zone?" To which we respond: *it depends*. It depends on your tolerance for risk, and how far over the edge you're willing to hang. It depends on your sense of urgency, and whether your goals can be reached in minutes or decades. And it depends on how reliably your preferred approach delivers actionable information and insight.

Those three parameters—risk, reward, and time—provide a way to compare nonordinary states. This sliding scale lets you assess otherwise-unrelated methods—from meditation to psychedelics to action sports, to any others you can think of. And you can distill these variables into an equation:

Value = Time × Reward/Risk

In this equation, *Time* refers to the learning curve, or how long you need to invest in a particular technique until it can reliably produce the experience of STER. *Reward* refers to how well we retain the insights that arise and how consistently they drive positive change. *Risk* refers to the potential dangers. If there's a chance that you could lose your life or your mind, that's something to consider well in advance. Put them all together and you get an approximate *Value* for each pursuit.

This calculus is reflected in the different treatments for PTSD we examined in the chapter on psychology. A one-day session with MDMA produces a marked decrease or abatement in symptoms, but you have to be willing to ingest an amphetamine to experience it. Five weeks of surfing—potentially less risky than a drug intervention—achieves a similar result, but entails learning a new sport in an unfamiliar and sometimes dangerous environment.

Meanwhile, meditation—both simpler and safer than surfing—requires twelve weeks and offers a slightly lessened benefit. These three approaches produce a similar reward (relief from trauma), but they come with varying degrees of risk and investment of time.

How you rank each variable is highly subjective—dependent on your abilities, responsibilities, and ambitions. But the final analysis is simple: are any of these pursuits worth the time, effort, and money we invest in them? Are we more energetic, empathetic, and ethical afterward? If not, they're just distractions or diversions from our lives. "I care not a whit for a man's religion," Abraham Lincoln once quipped, "unless his dog is the better for it."

And that goes double for techniques of ecstasy.

Hedonic Calendaring

In 1991, ARISE, the Associates for Research into the Science of Enjoyment, an organization that included representatives from the largest food and tobacco companies in the world, gathered in Venice, Italy. The meeting was dedicated to resisting "Calvinistic attacks on people who are obtaining pleasure without harming others." The topic of discussion: the Bliss Point.

The Bliss Point, as Pulitzer Prize–winning journalist Michael Moss describes in his book *Salt, Sugar, Fat,* is "the precise amount of sweetness [or saltiness or fattiness]—no more, no less—that makes food and drink most enjoyable." And not surprisingly, ARISE was deeply interested in hacking the Bliss Point.

"[Salt, sugar, and fat] were the three pillars of processed food, the creators of crave," Moss explains. "They were also the ingredients that more than any other were directly responsible for the obesity epidemic. . . . [T]he salt, which was processed in dozens of ways to maximize the jolt that tastebuds would feel with the very first bite . . . the fats, which delivered the biggest load of calories and worked more subtly in inducing people to overeat, and . . . the sugar, whose raw power in exciting the brain made it perhaps the most formidable ingredient of all."

For virtually all of evolutionary history, salts, sugars and fats were rare and precious. The only time we encountered sweetness was in the few weeks of berry season or the lucky find of a honeycomb. The phrase "worth his salt" refers to the days of Rome, when soldiers were paid in this essential mineral. And fat—concentrated, delicious calories—was only available in nuts, oils, and meats. This is why, when we encounter a bacon cheeseburger sandwiched between two Krispy Kreme donuts, we lose our minds.

In the quarter century since that Venice meeting, the food industry has become so good at manipulating these evolutionary impulses that we've eaten ourselves into a coma. According to the National Institutes of Health, 74 percent of American men and 33 percent of children are now overweight or obese. Our Bliss Point got hacked, and it's killing us.

Yet salt, sugar, fat provide a fraction of the payoff of ecstasis. In that state, we get access to all the brain's feel-good neurochemistry at once. For most of evolutionary history, nonordinary states were rare and precious experiences. So when we consider how readily accessible the four forces are making them today, it's important to remember that we're tinkering with impulses that are millions of years old. If the desire to get out of our heads is, as UCLA's Ron Siegel suggests, a "fourth evolutionary drive," right after food, water, and sex, then nearly unlimited access to ecstasis needs some checks and balances of its own. Otherwise our Bliss Point can become the flashpoint for a meltdown.

So how to pursue this path without getting "hooked on the high"? If we use the ecstasis equation to help us answer the question, "What is the *best* way to get into the zone?" then we need to add an additional concept here—hedonic calendaring—which helps us figure out *how often* we should get into the zone.

Hedonic calendaring provides a way to hack the ecstatic path without coming undone. It gives us a method to integrate hard-and-fast approaches like extreme skiing and psychedelics with slow and steady paths like meditation and yoga. It's one way to turn ecstasis into a sustainable long-term practice. And for anyone interested, there's a free downloadable Hedonic Calendaring PDF at www.stealingfirebook.com/downloads/

Step One: List everything you love to do (or that you'd like to do) that gets you out of your head. Action sports, yoga, live music, sex, brain stimulation, meditation, personal growth workshops, adventure travel, etc. This may seem simple, but if you consider the breadth of the Altered States Economy and the fact that we tend to compartmentalize the many different ways we modulate consciousness, putting it all down in one place can lend some fresh perspective.

Step Two: Use the Ecstasis Equation (Time X Reward/Risk) to rank this list for value. Think daily sun salutations versus an annual ultramarathon, or a ten-minute meditation versus a trip to see a Peruvian shaman.

Step Three: Sort your activities into one of five buckets: Daily, Weekly, Monthly, Seasonally, and Annually. More intense experiences typically provide more information but they do so at a higher level of risk. So it makes sense to allow plenty of time for recovery and integration between those sessions while committing to more frequent supportive practices.

How will you know if you've put the right activities in the right place? You'll feel mildly disappointed by how often you get to deploy a particular technique. When you deliberately combine ecstatic practices, you generate momentum surprisingly quickly and it's easy to feel like you're getting out of control. It's far better to start a little slowly, than skid off the track in turn one.

Step Four: Research shows we're more likely to keep habits that are tied to cultural milestones. So connecting practices to preexisting traditions can make them easier to stick to. Daily? Link it to sunrise or sunset, dinners, or bedtime. Weekly? Make it your own contemporary TGIF or Sabbath observance. Monthly? Connect it to the lunar cycle or the first or last days of the calendar. Seasonal? Solstices, equinoxes, Christmas, Easter, July Fourth, and Halloween all work and often come with vacation days attached. Annual? Take your pick: birthdays, anniversaries, New Year's, back to school, whatever's significant to you.

For the daily practices, you're trying to create self-propelling rituals so you will do them often enough to build a solid foundation. By hitching them to constant time slots and locations, you automate your positive behavior without having to draw down limited supplies of willpower. For the stickier (and likely, more enjoyable) weekly, monthly, and annual practices, you're putting in buffers to ensure you don't do them too much.

Step Five: Lastly, remember you're playing with addictive neurochemistry and deeply rooted evolutionary drivers. So, as your practices start building momentum, how do you know if you're pursuing a deliberate path or becoming a bliss junkie? Short answer? You don't.

Long answer: Once a year, set your indulgences up on a shelf, go thirty days cold turkey, and use this time to recalibrate. Attach the hiatus to traditional seasons of forbearance—Lent, Yom Kippur, Ramadan—or impose your own.

When you return to your practices, you've got perspective on how sticky things can get, and more feedback to fine-tune your calendar. Simply move any problematic activity to the right. If daily was too much, do it weekly. If quarterly practice leaves you wobbly, make it an annual event. Trade the morality of "should I or shouldn't I" for the logic of "more often" or "less often."

Altered states are an information technology and what you're after is quality data. If you spend all of your time blissed out, zenned out, drunk, stoned, sexed up, or anything else, then you've lost all the contrast that initially made those experiences so rich—what made them "altered" in the first place. By balancing inebriated abandon with monklike sobriety, ribald sexuality with introspective celibacy, and extreme risk-taking with cozy domesticity, you'll create more contrast and spot patterns sooner.

"The road of excess leads to the palace of wisdom," William Blake once wrote. Hedonic calendaring adds guardrails to that road. By dismantling the "oughts and "shoulds" of the orthodox approach, while avoiding the pitfalls of "if it feels good, do it" sensation seeking, we up the odds of getting to our destination in one piece.

There Is a Crack in Everything

If we can remember the known issues of STER, use the ecstasis equation to balance risk and reward, and deploy the hedonic calendar to avoid overheating, we should be good to go, right?

Maybe.

There's one final caveat worth keeping in mind. Namely, there's no escaping the human condition. We're born, we die, and figuring out the in between can be brutal. As Hemingway reminds us, "the world breaks *everyone*."

Yet so many ecstatic traditions have promised to repeal that fundamental law. If we can only unlock the secret, they say, we will have everything we want without any of the suffering. Ironically, the attempt to avoid suffering often creates more of it, leaving us susceptible to the most predictable trap of all: spiritual bypassing. "[It's] a widespread tendency to use spiritual ideas and practices," says John Welwood, the psychologist who coined the term, "to sidestep or avoid facing unresolved emotional issues, psychological wounds, and unfinished developmental tasks."

Typically, what gets bypassed on an ecstatic path are the mundane dissatisfactions of regular life. If those dissatisfactions are too intense, non-ordinary states can offer a tempting escape. But rather than bypassing these challenges, we can accept them and even draw power from them.

This response has a paradoxical name: vulnerable strength. Brené Brown, whose books and TED talks on the subject have resonated with massive audiences, explains it this way: "Embracing our vulnerabilities is risky but not nearly as dangerous as giving up on love and belonging and joy—the experiences that make us the most vulnerable. Only when we are brave enough to explore the darkness will we discover the infinite power of our light."

Balancing the bright lights of the ecstatic path with the darkness of the human condition is essential. Otherwise, we become unstable, top-heavy, our roots too shallow to ground us. The Indian philosopher Nisargadatta summed up the dilemma well: "Love tells me I am everything. Wisdom tells me I am nothing. And between these two banks, flows the river of my life." If we map this idea onto what we know about nonordinary states, then Nisargadatta's "everything/nothing" dialectic isn't just flowery wisdom, it's the by-product of the neurobiology of ecstasis itself.

The love that "tells me I am everything" arises from the awe and connection that we often experience in these states. Endorphins, oxytocin, and serotonin soothe our vigilance centers. We feel strong, safe, and secure. It's a welcome relief, and healing for those who don't often get to feel that way.

The wisdom that "tells me I am nothing" springs from the information richness. Dopamine, anandamide, and norepinephrine turn the bitstream of consciousness into a flood. Critical filters are down, pattern recognition is up. We make connections faster than we normally do. But within all that wisdom, there's a common tendency to be confronted by the hard truths we've been trying to ignore. "[Ecstasis] is absolutely ruthless and highly indifferent," wrote John Lilly. "It teaches its lessons whether you like them or not."

Every glimpse above the clouds can't help but suggest work still to be done on the ground. That's the resolution to the paradox of vulnerable strength. Ecstasis doesn't absolve us of our humanity. It connects us to it. It's *in* our brokenness, not in spite of our brokenness, that we discover what's possible.

The Japanese get at this same idea with the concept of *wabi sabi*—or the ability to find beauty in imperfection. If a vase is accidentally broken, for example, they don't throw the pieces away or try to patch it up to hide the accident. Instead, they take golden glue and painstakingly reassemble the vessel, so its unique flaws make the piece more beautiful.

The late poet and musician Leonard Cohen may have been our greatest contemporary commentator on this theme. In his song "Anthem," he sings: "Ring the bells that still can ring, forget your perfect offering. There is a crack, a crack in *everything*. It's where the light gets in."

The ecstasy will always come with the agony—that's the human condition. Nothing we do absolves us from the broke-open beauty of that journey. So there will be cracks. Thankfully, there will be always be cracks. Because, as Cohen reminds us, that's where the light gets in.

CONCLUSION

Row Your Boat or *Fly* Your Boat?

In 2013, what Larry Ellison wanted, beyond all reason, was a win. So the founder of the software giant Oracle and one of the richest men in the world spent more than $10 million building the fastest boat ever to compete in the America's Cup. The Oracle catamaran was equipped with futuristic hydrofoils that lifted the entire boat out of the water, enabling speeds of up to 55 knots. It defied all prior limits of wind-powered watercraft.

But none of that technology helped the Oracle team beat the New Zealand boat in the first six races of the finals. Despite assembling a collection of software engineers to map and plan every detail, Ellison's team was only one heat away from getting trounced by the upstart Kiwis. The races just weren't playing out the way they had in their complex computer models.

In fact, all of that expensive technology obscured what the skipper, Jimmy Spithill, knew in his bones—the New Zealand boat was killing them by sailing a radically different course. One of the hardest things to do efficiently in sailing is head into the direction the wind is coming from. It's easy to get blown downwind—anyone with a canoe and a large trash bag can pull that off. But heading upwind requires precisely balancing the force in the sails against the resistance provided by the daggerboards (the large fins under the boat).

Rather than the standard "high and slow" zigzag of the Oracle team, where they sailed about 45 degrees into the wind, the Kiwis had broken with convention and were sailing "low and fast"—50 or even 60 degrees off the direction of the wind. By doing that, their boat was able to pop up on its hydrofoils and rip along nearly 30 percent faster than their competition. Sure, they had to cover a little more ground, but they were doing it at warp speed.

Down six races and with nothing left to lose, Spithill defied Ellison and fired his tactician. For the remainder of the finals, he sacrificed precious upwind headway to capitalize on the efficiencies that the hydrofoils afforded. Once he threw out the old conventions and adapted to the true potential of his boat, he skippered Oracle to eight straight wins and the greatest comeback in the oldest international sporting event in history.

A similar choice faces us today. Just as adding hydrofoils to the Cup boats changed what was possible on (and just above) the water, unprecedented access to ecstasis has changed what is possible in our lives. Experiencing the selflessness, timelessness, effortlessness, and richness of nonordinary states of consciousness can accelerate learning, facilitate healing, and provide measurable impact in our lives and work. But we have to revise our tactics and upend convention to make the most of those advantages.

Just as old sailing wisdom favored "high and slow"—meaning that you pointed your boat as close to the eventual upwind destination as possible—we are steeped in a "high and slow" culture of relentless goal setting and linear forward progress. It's why, in the United States, more than half of paid vacation days go unclaimed and we perversely brag about clocking 60–80-hour workweeks (even though our effectiveness drops after 50 hours). We valorize suffering and sacrifice, even when the victories they provide are hollow.

Surrendering any of that hard-fought ground to pursue nonordinary states can seem, at first glance, irresponsible, or, at a minimum, deeply counterintuitive. Spithill learned that the performance gains of hydrofoiling were so pronounced that if he didn't change how he navigated, he'd lose to those who did.

The same is true for ecstasis. Research shows that these experiences lift us above normal awareness, and propel us further faster. Much of our conventional schooling, personal development, and professional training still miscalculate this fact. It's hard to fathom how much faster we can go, how much more ground we can cover, if we can only appreciate what high performance now looks like.

An Afterthought

While most schoolchildren can recall the broad outlines of the story of Prometheus stealing fire and getting chained to a rock, few of us know why he did it, what came next, or what it might tell us about the road ahead.

Back in mythic prehistory, Zeus hadn't always been king. He'd had to battle a race of giants, the Titans, to claim his throne on Olympus. And when he won, he banished all of them, except for two young brothers, Prometheus and Epimetheus, whom he charged with making all the living creatures on the earth.

Epimetheus, whose name means "afterthought," started making animals out of river clay and bestowing on them all the gifts that Zeus had allotted: strength, speed, fangs, and fur. But he worked so hastily that by the time Prometheus had finished fashioning humans into the likeness of gods, there were no more advantages to bestow. It was then that Prometheus, whose name means "forethought," took pity on man, shivering and defenseless in the dark, and asked Zeus if he could give them fire to compensate. Zeus said no, Prometheus stole fire anyway, and got punished. That's the part we all remember. But Zeus wasn't finished with the humans, or the brothers.

He wanted to make sure that no one challenged his power

ever again. So he made a woman, Pandora, whose name means "all giving," and gave her a box filled with the tragedies of life to unleash on the world. Prometheus, true to his name, was cautious about accepting a gift from Zeus, but the more impulsive Epimetheus fell hard for the beautiful girl and they married.

Eventually—as Zeus knew she would—Pandora gave in to her curiosity and opened the box. When she did, all the scourges of existence—war, pestilence, famine, greed—flew out to torment mankind. At the last minute, though, Pandora managed to snap the lid shut, leaving one thing remaining: hope. "That is why," explained the Greek storyteller Aesop, "hope alone is still found among the people, promising that she will bestow on each of us the good things that have gone away."

Only now, our hope isn't as blind as it used to be. We don't have to keep making sacrifices to powers beyond our understanding or keep waiting for rescue from our plight. Today, a string of modern-day Prometheans have taken up the torch. Rather than relying on a singular emissary, we can draw information and inspiration from a global network of pioneers and innovators. And this ups our odds considerably. From 3D chem printers allowing us to explore our minds, to full-spectrum sexuality letting us lose ourselves in each other, to transformative technologies nudging us into clearer self-awareness, to giant gatherings providing a taste of communitas, the four forces have unlocked the gates of Olympus. And this may offer the greatest hope of all. We no longer have to rely on someone stealing fire for us.

Finally, we can kindle that flame ourselves.

If this book has lit a fire and you're interested in learning more, join the conversation. We've got free tools to tally your own Altered States Economy, plan your Hedonic Calendar and discover your flow profile. We also offer intensive trainings to unlock personal and organizational high performance. And if you just want to reach out, share your stories, ask some follow up questions, or let us know where we're completely out to lunch, that's cool too.

www.stealingfirebook.com

ACKNOWLEDGMENTS

Stealing Fire has benefited greatly from the insight and generosity of many: our agent, Paul Bresnick, our editor Julia Cheiffetz, and the great team at Dey Street and Harper Collins. Michael Wharton was a tireless warrior. A great many other friends and colleagues as well: Jason Silva, Salim Ismail, David Eagleman, Judson Brewer, Andrew Newberg, Bill Tai, Suzy Mai, Jeff Rosenthal and the entire Summit Crew, Kristen Ulmer, Dean Potter, Matt Reardon and the Squaw Valley posse, everybody who attended any of FGP's Stealing Fire events at Esalen, the amazing staff at Esalen, our ninja assistant Lucas Cohen, Dave Asprey, the great team at Google, but especially Adam Leonard and Anthony Slater, Tim Ferriss, Michael McCullough and the Stanford Brain-Mind Conference, the U.S. Naval War College, the leadership of DEVGRU, SEALFIT founder Mark Divine, Burning Man CEO Marian Goodell, James Hanusa, Android Jones and the entire global community of Burners, Chip Conley, Mike Gervais, Ken Jordan and Evolver, Rick Doblin and MAPS, Nicole Daedone, Neil Strauss, Laird and Gabby Hamilton, Jimmy Chin, Hollis Carter, Michael Lovitch and the BabyBathwater Crew, Colin Guinn and the team at Hangar, Mikey Siegel and the Transformative Tech tribe, James Valentine, Richard Branson for hosting us on Necker, Sergey Brin, Larry Page, Astro Teller, Bob Kegan, Robin Carhart-Harris, Skip Risso and the USC Institute for Creative Technologies, Gordon and Courtney Gould at Smarty Pants, Lashaun Dale, Claudia Welss, Chris Berka and her team at Advanced Brain Monitoring, Laura Anne Edwards, Shahar Arzy, Molly Crockett, Andrew Hessel, and Bob Coyne for his kickass book cover. Most important, our wives and families, who made all of this possible.

A QUICK NOTE ON INSIDE BASEBALL

In this book, we've attempted to stitch together a far-ranging argument across many distinct disciplines. So we've had to make continuous choices as to how deeply we dive into each subfield and how to distill the most relevant findings.

As a result, we've deliberately steered clear of "inside baseball"—meaning, those details, discussions, and disputes that endlessly engage specialists and would needlessly confuse the generalist. Where there's a diverging consensus in a certain field of study, we have tried to walk back the discussion to the last prior solid ground. Where there's no solid ground, we've chosen what we feel to be the most credible or interesting to report.

What follows is a listing of some of the more obvious inside conversations we've stayed away from. If you are passionate about one or more of these areas, we've run the distinct risk of underwhelming you in our treatment, or leaving ourselves open to a technician's critique. We hope that the bigger story we're telling still occasionally surprises and mostly satisfies.

In approximate order in how they appear in the book:

Ingredients for Kykeon: Ergot is a less than satisfactory final solution to *kykeon,* but has yet to be definitively replaced with another clear favorite. We do know that *kykeon* was required to be diluted "ten parts to one with regular wine" which suggests it was potent, and we do know it was enjoyable enough to steal for a party. Beyond that, it's very hard to separate out the "set and setting" of the entire acculturated ritual of Eleusis from the specific impact of the substance itself. As with decoding the Hindu Soma, it would unlock a fascinating chapter in religious and social history to be able to cross-reference specific chemical compounds with the philosophy, art, and politics they inspired.

SEALs and the Halo Effect: In the past fifteen years, special operations forces in general and the Navy SEALs in particular have borne a disproportionate brunt of both tactical responsibility on multi-front wars and media scrutiny at home. As a result, they have enjoyed both uncritical adulation and increasingly critical assessments (the *New York Times,* in particular has published several extended investigative pieces on serious topics involving DEVGRU). In our coverage, we have reported on the teams themselves and our experience of them, their philosophy and training methods. There are other conversations to be had, in how the teams are deployed and specific lapses of judgment or errors in execution, but we stand by our respect for what they do and how they do it.

fMRI Reliability: In June 2016, the *Proceedings of the National Academy of Sciences* published a paper pointing out serious flaws in the most widely used software algorithms that decode fMRI data, and that these algorithms might have resulted in false positives up to 70 percent of the time. If true, this would largely invalidate most studies conducted before 2015 (including several that we reference in this book). Furthermore, there are those in the research community who believe that fMRIs have been overused and question the correlation between blood flow to a brain region (what fMRIs most readily measure) and thinking or activity in that region (especially since younger children can demonstrate mental activity without any blood flow to the supposedly involved regions). So, expect a lot of studies to be rerun, and many revised over the coming few years. While the details are subject to constant revision, we believe that the fundamental argument we're making about our neurobiological "knobs and levers" affecting our psychological experience will only become stronger over time.

Transient Hypofrontality and Selflessness: In writing this book, we set out to distill all the current theories of selflessness into one coherent meta-argument. Despite interviewing many of the major players in this field, we couldn't quite get to "the simplicity on the other side of complexity." So, for now, the leading candidates to explain how our waking conscious self shuts off in nonordinary states are: transient hypofrontality, transient hyperconnectivity, default mode network interruption, and cortico-thalamic gating (Henri Bergson's original idea, which Aldous Huxley popularized in *The Doors of Perception*). We suspect that several of these may well work in sequence or concert, or may produce similar experiences of selflessness via different mechanisms of action, and a few may be disproved or revised significantly. We also suspect that as measurement devices get more sophisticated at capturing real-time diachronic activity (as opposed to static snapshots) we'll get a much fuller-fledged picture that integrates neuroanatomy, neuroelectricity, and neurochemical interactions in a coherent model of selflessness. Until that time, we have presented some of the findings that seem most credible and relevant to our study of ecstasis, while leaving lots of room for updates. (For researchers interested in advancing this research please contact us at info@flowgenomeproject.com.)

SSRIs and Medicating Psychology: In general, we support the appropriate and selective use of psychopharmaceuticals under the supervision of medical practitioners and we acknowledge they can be life-changing under the right conditions. We are also critical of overprescription, undersupervision, and pervasive off-label uses where apparent market forces are trumping more effective (and often, nondrug) interventions.

Perils of Porn: In the past few years there has been a rising tide of social commentary against the downsides of online pornography—ranging from marital troubles, teenage sexual norming, erectile dysfunction, and outright addiction. More recently, there have been studies refuting some of those claims, suggesting that porn viewing has fewer of those adverse effects and that most viewers actually experience an increase in face-to-face intimacy and arousal. We do not expect this debate to be resolved anytime soon, and have instead chosen to focus on the simple observation that many people are engaging it, and therefore it's worthy of consideration as an indicator of social trends and the altered states economy.

Psychedelic Renaissance: There has been exhaustive coverage in the past several years of the new rounds of studies involving psilocybin, LSD, MDMA, marijuana, and ayahuasca. We have not attempted a comparable survey here, believing it to be thoroughly addressed in other treatments. We have chosen instead to contextualize this research within the bigger conversation of ecstasis, the measurable benefits of momentary selflessness, and the case studies this research is providing.

Selfishness of Extreme Athletes: Every time an extreme athlete dies, whether it's Shane McConkey or Dean Potter, or any of the dozens of leaders in the outdoor community who perish annually, people invariably chime in with how selfish they were to risk their lives on elective pursuits. How could they, the critics ask, leave behind grieving spouses, children, and even pets (in the case of Dean's dog Whisper). As with our treatment of pornography and psychedelics, we have chosen to steer clear of moralizing others' choices and instead focus on the simple fact that athletes continue to risk their lives, and presumably do it for a series of more or less considered reasons. We want to get as close as we can to understanding those reasons and connect them to the larger draw and benefit of ecstatic states. And, for what it's worth, we believe there is a comparable, though far less sensational risk to living a suburban life of quiet desperation and dying prematurely of a lifestyle disease without having once tasted what these athletes always live (and only sometimes die) for.

Overhyped Sensor Tech: In several places, we highlight the accelerating potential of smart sensors and wearables to give us more feedback on our bodies and brains. In the past few years, the hype cycle has claimed several high-profile victims as the Federal Trade Commission, Food and Drug Administration, and class-action suits have clamped down on Nike, Apple, and Lumosity for claiming benefits or accuracy they could not deliver. In our research, we generally break down wearable tech into three buckets. First, the DARPA-level bleeding-edge tools that cost tens of thousands to millions to make, are not federally approved, and generally need an overseeing physician in the context of a formal study to deploy. Then there are the "prosumer" grade devices that cost thousands to tens of thousands, provide near-research-grade data, and can be deployed by skilled professional practitioners or well-funded biohackers. Finally, there's the consumer-grade tools, available for under one thousand dollars and often for little more than the cost of an app plus a peripheral device. These are worthwhile mostly as "edutainment," and as the lawsuits and federal interventions suggest, are not always all they're cracked up to be. That said, even these entry level tools are getting better by the year and will likely follow the Moore's law trends we see in most tech fields.

Optimal Psychology: As with the Psychedelic Renaissance, this is a field amply and ably covered by others. And while we celebrate the move toward the study of wholeness and happiness, most of the field is dedicated to the incremental movement to make us "10 percent happier." Given that the percentage gains in performance from ecstasis range from 200 to 500 percent, we are more interested in higher-impact options (while also supporting all efforts to build on-ramps to these experiences).

Burning Man Elitism: Yup. Lots of it, likely getting worse. But we believe there's still much more to consider and contextualize than class warfare. After all, Aspen, Colorado, started out as a blue-collar mining town before becoming a countercultural haven before becoming the LA-NY annex it is today. Doesn't make those mountains any less beautiful, though. What remains most interesting to us is considering the event as a "sandbox for the future," where we can spot many of the forces of ecstasis out of the lab and in their natural habitats. When we discuss the influence of Burning Man culture on the social elite, we are more interested in what's different and novel these days—specifically the three-part notion of how attendees have a disproportionate influence on media and markets, how its members are extending into some of the more conventional and powerful organizations around the world, and how practical innovations are showing up around the world.

Our Own Inside Baseball: Throughout this book, we tell the stories of leaders in fields—many of whom, in the course of our research, we've come to know and like. They are innovators in the domains that fascinate us, and we often share similar orientations. Some, like DEVGRU commander Rich Davis and Advanced Brain Monitoring CEO Chris Berka, we've become friendly colleagues with, swapping notes and ideas several times a year. Others, like philosopher Jason Silva, Red Bull's Andy Walshe, kitesurfers Bill Tai and Susi Mai, filmmaker Jimmy Chin, neuroscientist Andy Newberg, late base jumper Dean Potter, and skier Kristen Ulmer, we've asked to serve on the advisory board of the Flow Genome Project to help us advance the mission of open-sourcing the science of peak performance. These are volunteer, unpaid positions. Furthermore, their accomplishments precede their positions in our organization and will continue long past that service ends, and we are grateful to know them and have their support. (And you all will enjoy hearing of their accomplishments; they are all remarkable.)

NOTES

Introduction: The Never-Ending Story

1 This one kicked off: E. F. Benson and Craig Peterson, *The Life of Alcibiades* (N.p.: CreateSpace Independent Publishing Platform, 2010), pp. 127–38. Also see Robert Strassler, ed., *The Landmark Thucydides: A Comprehensive Guide to the Peloponnesian War* (New York: Free Press, 1996).

1 Alcibiades, a prominent Greek general: Plutarch, *Lives IV, Alcibiades and Coriolanus. Lysander and Sulla* (Cambridge, MA: Harvard University Press, 1916).

1 "In less than an hour": Plutarch, *On the Soul*, quoted in Stobaeus, IV, as translated by George E. Mylonas, *Eleusis and the Eleusinian Mysteries* (Princeton, NJ: Princeton University Press, 1961), pp. 246–65. Also see Edith Hamilton, *The Greek Way* (New York: Norton, 1993), p. 179.

2 "And not just any mysteries; the Eleusinian Mysteries": Will Durant, *The Life of Greece* (New York: Fine Communications, 1997), pp. 188–93. Also, IO9 did a great little article on the mysteries. See http://io9.gizmodo.com/5883394/the-eleusinian-mysteries-the-1-fraternity-in-greco-roman-society.

2 "Our Mysteries had a very real meaning": Plato, *Phaedo,* translated by F. J. Church (London: Pearson, 1987), p. 69.

2 Cicero went further:" Cicero, *On the Laws,* translated by C. W. Keyes (Cambridge, MA: Harvard University Press, 1928), p. 14.

3 Swiss chemist Albert Hofmann: R. Gordon Wasson, Albert Hofmann, and Carl Ruck, *The Road to Eleusis* (Berkeley, CA: North Atlantic Books, 2008).

3 When consumed accidentally: Ibid.

3 At the center of this dynamic sits the myth of Prometheus: Robert Graves, *The Greek Myths* (Mount Kisco, NY: Moyer Bell, 1955).

4 at the Flow Genome Project": See www.flowgenomeproject.com.

5 Scientists have known about the relationship: For a full breakdown of the history of flow science, see Steven Kotler, *The Rise of Superman* (Boston: New Harvest, 2014).

5 clandestine experiments with "ecstatic technologies": The term "ecstatic technologies" is an update on University of Chicago historian Mircea Eliade's classic description, *Archaic Techniques of Ecstasy*." Eliade uses the term to refer to all the original ways shamans altered their consciousness. We have replaced "techniques" with "technologies" because of the range of state-changing devices now available, including things like neurofeedback, isolation float tanks, transcranial magnetic stimulation, and other recent additions to this canon.

6 This journey has led us all over the world: Ok—here's the deal, included in this list are two stories didn't make it into the book for editorial reasons, but we wanted to share them nonetheless.

A few years ago, we were invited to Moscow to advise an incredibly progressive $100M education company founded by a former Communist Youth Party leader (he explained that back in those days, the only way to get out from under the thumb of the party was to lead it). We ate dinner at midnight in Peter and the Wolf style restaurants that were favorites of the Politburo, and routinely worked until 4 am. But our hosts also shared that much of their inspiration came from regular ayahuasca ceremonies they had their whole company participate in (not always the wisest idea!). In fact, they had a favorite Peruvian shaman whom they flew all around the world to meet them in holy sites and conduct ceremonies at various times of the year. Their most outrageous stunt, was pulling strings with Egyptian government officials so they could sneak into the Great Pyramid of Giza and host an "aya" ceremony in its exact center at the Vernal Equinox.

In the summer of 2016 we were invited to speak at the United Nations HQ in New York to a global gathering of young leaders and entrepreneurs on the role of Flow in social change. The night before, we attended a private dinner on the role of entheogenic plants in leadership, government and policy change. The entire gathering was conducted under "Chatham House Rules" meaning participants could act on information shared, but not identify who said/did what. To that end, everyone at the table was exploring psychedelics and then applying the insights to affect NGO and governmental work in Caribbean nations, Asian economic development, and Amazonian reforestation efforts. Not coincidentally, two-thirds of the gathering were also meeting back up the next month at Burning Man.

Part One: The Case for Ecstasis

7 The alternative is unconsciousness: David Foster Wallace, *This Is Water* (New York: Little, Brown, 2009), p. 123.

Chapter One: What Is This Fire?

9 One of the hardest parts of being a Navy SEAL: Author interview with Rich Davis, 2013.

9 That's what made capturing Al-Wazu: At the request of DEVGRU and for reasons of security, "Al Wazu" is not his real name. Also, the story of the capture of Al Wazu was told to us by Commander Davis, then fact-checked with the Navy's legal department (as much of the actual story remains classified).

11 The Greeks had a word: Obviously you can find translations of this word everywhere, but author Jules Evans does a particularly good job explaining it on his website, Philosophy for Life," writing: "Ecstasy comes from the ancient Greek exstasis, which literally means 'standing outside,' and more figuratively means 'to be outside of where you usually are.' In Greek philosophy, in Plato and Neoplatonists like Plotinus, it came to mean moments when a door opens in your mind or soul, you feel an expanded sense of being, an intense feeling of joy or euphoria, and you feel connected to a spirit or God. Its closely connected to another word in Plato, *enthousiasmos*, which means 'the God within.' So in moments of ecstasy, according to Plato, you stand outside of yourself, and God appears within you." See http://www.philosophyforlife.org/modern-ecstasy-or-the-art-of-losing-control/.

11 Keith Sawyer in his book *Group Genius*: Keith Sawyer, *Group Genius* (Basic Books, 2008), p. 4.

12 It costs $25,000 to turn: There are a number of different ways to calculate this number and we blended a few different stats to come up with our total. See Diana Olick, "An Army of One Carries a High Price," NBCNews.com, October 21, 2002, available at http://www.nbcnews.com/id/3072945/t/army-one-carries-high-price/#.V-0OhDKZO50. Also see Jared Hansbrough, "An Activity-Based Cost Analysis of Recruit Training Operations at Marine Corps Recruit Depot, San Diego, California," Naval Postgraduate School, Monterey, CA, 2000.

12 Estimates for eight weeks: Stephanie Gaskell, "Three Navy SEALs Freed Capt. Phillips from Pirates with Simultaneous Shots from 100 Feet Away," New York *Daily News,* April 14, 2009, http://www.nydailynews.com/news/world/navy-seals-freed-capt-phillips-pirates-simultaneous-shots-100-feet-article-1.360392.

12 As it costs about $1 million a year": Thomas Smith, "Money for American Commandos," *Human Events,* April 23, 2008.

12 As their official website explains: See https://www.sealswcc.com/navy-seals-benefits.html.

13 As SEALFit founder Mark Divine: T. J. Murphy, "The World's Most Intense Fitness Program," *Outside*, December 16, 2014.

14 The Navy's caste system: Richard Marcinko, *Rogue Warrior* (New York: Pocket Books, 1993), p. 8.

17 "I do not seek recognition": Posted by Mark Divine, "The Navy Seal Code," NavySeals.com, 2016.

18 "Larry and I [had] managed": Gregory Fernstein, "How CEOs Do Burning Man," *Fast Company*, August 27, 2013.

18 . . . *New York Times*' John Markoff's assessment": John Markoff, "In Searching the Web, Google Finds Riches," *New York Times,* April 13, 2003.

19 The company that set the bar: For a complete breakdown of Google's involvement at Burning Man, see Fred Turner, "Burning Man at Google," *New Media & Society* 11 (2009): 73–94.

20 Eric was the only one: Gregory Ferenstein, "How CEOs Do Burning Man," *Fast Company*, August 27, 2013; John Markoff, "In Searching the Web, Google Finds Riches," *New York Times,* April 13, 2003; and the original citation, Doc Searls, Harvard Berkman fellow, 2002, http://doc.weblogs.com/2002/12/10.

20 Stanford sociologist Fred Turner: Turner, "Burning Man at Google."

20 Attending festivals like Burning Man: Author interview with Molly Crockett, 2016.

20 In those states, all of the neurochemicals: For a complete breakdown of the neurochemistry of flow, see Kotler, *The Rise of Superman* (Boston: New Harvest, 2013), pp. 65–69.

21 The whole point of taking Schmidt to Burning Man: Author interview with Salim Ismail, 2016.

21 When Google hired Schmidt: Markoff, "In Searching the Web, Google Finds Riches"; Quentin Hardy, "All Eyes on Google," *Forbes*, May 26, 2003. Revenue numbers were privately held for Google in 2001 but were reported at around $100 million, tripling the following year to $300 million. By 2011, when Schmidt stepped down, revenues were $37.9 billion.

23 A very specific range of nonordinary states of consciousness (NOSC): Walter Mead interviewing Stan Grof, "The Healing Potential of Non-Ordinary States of Consciousness," see: http://www.stanislavgrof.com/wp-content/uploads/2015/02/Healing-Potential-of-NOS_Grof.pdf

24 Regular waking consciousness has a predictable and consistent signature: For a look at the absolute basics of consciousness, See Christof Koch, Marcello Massimini, Melanie Boly, and Giulio Tononi, "Neural Correlates of Consciousness: Progress and Problems," *Nature Reviews Neuroscience* 17 (2016): 307–21. Also see Angela Clow, Frank Hucklebridge, Tobias Stadler, Phil Evans, and Lisa Thorn, "The Cortisol Awakening Response: More than a Measure of HPA," *Neuroscience and Biobehavioral Reviews* (2010); doi:10.1016/j.neubiorev.2009.12.011.

24 During the states we're describing: There's obviously a ton of information on this one, but especially relevant are Arne Dietrich, "Functional Neuroanatomy of Altered States of Consciousness," *Conscious Cognition*, June 12, 2003, pp. 231–56; Matthieu Ricard and Richard Davidson, "Neuroscience Reveals the Secret of Meditation's Benefit," *Scientific American*, November 1, 2014; J. Allan Hobson, *The Dream Drugstore* (Cambridge, MA: MIT Press, 2001); Kotler, *The Rise of Superman*; and C. Robert Cloninger, *Feeling Good* (Oxford: Oxford University Press, 2004). Also see Arne Dietrich's excellent TED talk, "Surfing the Stream of Consciousness," https://www.youtube.com/watch?v=syfalikXBLA.

24 The knobs and levers being tweaked in the brain: See www.flowgenomeproject.com/stealingfiretools.

26 "We're a very high-performing club": Author interview with Rich Davis and other team leaders (who have likewise needed to remain anonymous), 2013.

26 researcher and neuroscientist John Lilly: Float tank histories are everywhere, but to hear John Lilly tell it, see John Lilly, *The Scientist* (Berkeley, CA: Ronin, 1996), pp. 98–108.

27 "It's going well": Author interview with Adam Leonard, 2013.

28 Unlike those of many other firms: Turner, "Burning Man at Google," p. 78.

28 The Altered States Economy: In putting this together, we used the '"knobs and levers" of ecstasis as our guide. If a particular item in our tally pushed brainwaves out of the normal beta band of waking consciousness, triggered the release of at least two of the big six state-shifting neurochemicals (dopamine, norepinephrine, anandamide, serotonin, endorphins and oxytocin), or deactivated/hyperactivated the prefrontal cortex and/or the default mode network to produce the experience of STER, we included it in our tally.

We decided to focus our attention on four main categories: Drugs, Therapy, Media and Recreation. In assembling our numbers, we also took the most recent data available, but made no attempt to translate these figures into 2016 dollars. This means most of our global numbers are probably lower than they would be today. Similarly, in a few instances, only U.S. numbers were available. In those cases, we took only the U.S. numbers and did not attempt to scale things up for the rest of the world because there are asymmetrical adoption of many of these economies—especially personal growth, action sports, etc. (unless otherwise noted, all numbers reflect global numbers). That makes this estimate initial and rough, in the spirit of Enrico Fermi's guesstimates. We encourage scholars who'd like to take a crack at a more refined tally to do so, updating the ASE to global numbers in current-year adjusted dollars. And let us know, at flowgenomeproject.com/stealingfiretools.

In our accounting, we started with the assumption that the price tag of ecstasis should include all psychoactive pharmaceuticals. We didn't consider drugs intended to address a physiological conditions—like blood pressure, or cholesterol—only those that specifically targeted and shifted states of consciousness in the user. That meant the broadly prescribed pain, depression, anxiety, attention, and insomnia meds, along with long-tail drugs used to manage more serious mental disorders. This totaled out to $182.2 billion.

See for mental health drugs: BCC Research, "Drugs for Treating Mental Disorders: Technologies and Global Markets," January 2011, PM074A. For prescription pain management drugs, see BCC Research, "The Global Market for Pain Management Drugs and Devices," January 2013, HLCO26D. For sleep aids: BCC Research, "Sleep Aids: Technological and Global Markets," June 2014, HLCO81B.

We also included all licit and illicit substances that helped people get out of their heads"—from alcohol and tobacco and marijuana on the licit side to cocaine, heroin, methamphetamines, and marijuana on the illicit side. We further included all of the prescription drugs of abuse (like Ritalin and OxyContin) that are prevalently used off-prescription or resold on gray and black markets. We also focused only on the production and consumption of these substances and didn't take into account any collateral costs, like law enforcement, incarceration, political lobbying, and judicial processes. This added up to roughly $2.3 trillion.

For illegal drugs, see United Nations Office on Drugs and Crime, "World Drug Report 2005," vol. 1, Analysis. For marijuana, see Arcview Market Research, "The State of Legal Marijuana Markets," 4th ed., 2016; "Alcohol: Research and Markets, Global Alcoholic Beverage Industry-Forecast to 2018," February 21, 2014. For coffee: Wevio, "Global Coffee Industry Facts & Statistics of 2014–2015," May 29, 2015. For tea: Transparency Market Research, "Tea Market-Global Industry Analysis, Trend, Size, Share and Forecast 2014–2020," March 3, 2015. For energy drinks: "Global Energy Drinks Market: Insights, Market Size, Share, Growth, Trends Analysis and Forecast to 2021," April 2015, http://www.researchandmarkets.com/reports/3161745/global-energy-drinks-market-insights-market. For tobacco: Euromonitor International, "Global Tobacco: Findings Part 1—Tobacco Overview, Cigarettes and the Future," June 2014.

Then we widened our net and tried to include all of those goods and services expressly dedicated to solving the same dilemma—how do I get out of my current state of awareness. How do I shake a lived experience that ranges from unsatisfying to insufferable? Enter the blended category of neurotech, psychology, psychiatry, counseling, and self-help. Here we wanted to include all therapies, personal development, book sales, workshops, and information trainings (discounting those geared toward some professional skill acquisition and focusing solely on those that promise some change of existential perspective—aka Help Me Get Happier!). Unfortunately, most of those numbers are not available. What we could get was the $$3.65 billion in neuromodulation (that is tech designed to shift states), the $11 billion Americans spend on self-help annually, the $869 million Americans spend on life coaches, the $2 billion spent on business coaches (of which we counted only 50 percent, assuming the other portion related to skills acquisition), and the $15 billion Americans spent on therapy and counseling. For certain, this number is significantly less than the actual size of the global market, but this totaled out to $31.52 billion.

For neuromodulation tech: "Markets and Markets, Neuromodulation Market by Technology, by Application-Trends and Global Forecasts to 2020." For self-help: Marketdata Enterprises, "The U.S. Market for Self-Improvement Products & Services," November 2010. For personal coaches: IBISWorld, "Life Coaches in the US: Market Research Report," February 2016. For psychology and counseling: IBISWorld, "Psychologists, Social Workers & Marriage Counselors in the US: Market Research Report," February 2016. For business coaches: "Inside the Coaching Industry," *Success,* June 30, 2015.

Next, we turned our attention to media. We started with the video game industry. The combination of big screens and increasingly complex character-driven story lines has made video games so out-of-our-heads immersive that the industry has become bigger in size and value than the movie industry. More specifically for our purposes, games are built around a number of core flow triggers, including novelty, complexity, unpredictability, and very precise challenge and reward ratios. Add in another $99.6 billion. See Newzoo, "Global Games Market Report," April 21, 2016.

Virtual reality, which is specifically designed to be far more immersive than video games, also counts. What's more, the extra burst of immersion makes VR exceptionally good at pulling more of flow's triggers. So we added another $12.1 billion to our total. See "Virtual Reality Market Size Worldwide 2016–2020," 2016, https://www.statista.com/statistics/528779/virtual-reality-market-size-worldwide/.

We also took into account the social media market, estimated at $17.2 million in 2015. See http://trade.gov/topmarkets/pdf/Media_and_Entertainment_Top_Markets_Report.pdf.

Out of the rest of more traditional film and TV, you can make the case that with the exception of documentaries (where we might actually be trying to learn something), virtually all visual storytelling serves to provide an escape. And there's a reason for this. Our brains do a pretty crummy job of distinguishing the physical from the filmed. For 99.9 percent of human evolution, if we saw something with our eyes, that meant it was in our physical world. The brain regions for fear, love, and movement, along with our mirror neurons, spike our ability to imagine ourselves right up there with our screen heroes. The effect is, quite literally, intoxicating.

But again we wanted to err on the side of conservative. So we limited our accounting to two categories of screen viewing that are especially and purposely escapist: IMAX/3D films (where the record-setting film *Avatar* serves as the exemplar) and binge watching of streaming shows (where Netflix's *House of Cards* practically invented the term). We thus get a good sample of media consumption specifically designed to prompt a loss of self. In the case of IMAX, revenues hit $1 billion. See "IMAX Corporation Reports Fourth Quarter and Full Year 2015 Financial Results," Imax.com, February 24, 2016, http://www.imax.com/content/imax-corporation-reports-fourth-quarter-and-full-year-2015-financial-results.

Add to that the more recent phenomenon of binge-watching entire TV seasons in one sitting, where the magic of "maybe" dopamine makes season-ending cliffhangers compel otherwise reasonable couples to stay up way past their bedtime. This number is tricky to come by, but a recently released Deloitte study 70 percent of TV watching is binge watching, so we took 70 percent of the digital streaming market size, or $21.20 billion.

For binging overview: Claire Gordon, "Binge Watching Is the New Normal," *Fortune,* March 24, 2016. Also, Digital Democracy Survey, fielded in 2014, https://www2.deloitte.com/content/dam/Deloitte/global/Documents/Technology-Media-Telecommunications/gx-tmt-deloitte-democracy-survey.pdf. For the size of video-streaming market: "Markets and Markets, Video Streaming Market by Streaming Type (Live Video Streaming and Non-Linear Video Streaming), by Solution, by Service, by Platform, by User Type, by Deployment Type, by Revenue Model, by Industry, and by Region—Global Forecast to 2021," May 2016, http://www.marketsandmarkets.com/Market-Reports/video-streaming-market-181135120.html.

We also added in the $6.2 billion EDM industry and the $97 billion porn industry. For EDM: Kevin Watson, "A Study of the Regional Electronic Music Industry," *IMS Business Report 2015: North America Edition,* p. 12. For porn, see Chris Morrow, "Things Are Looking Up in America's Porn Industry," CNBC, January 20, 2015.

There's also the realm of recreation: those pastimes, sports, and activities whose central function is to get us out of ourselves. Certainly, low-skills/high-thrills activities like bungee jumping and roller coasters qualify—for those moments of self-inflicted terror we are most definitely not worrying about our taxes, or our last breakup. So too with action and adventure sports where the rush, the stoke," is the thing that drives people to come back for more. Here's we'll discount bat and ball and team sports and focus on those "gravity games" with strong movement and high consequence, like skiing, climbing, mountain biking (classic triggers for driving people into what researchers call the deep now). We also included the $27 billion yoga industry and the $1 billion meditation. Lastly, we included the category of '"adventure travel,'" where the rigors of the journey are all but guaranteed to get you out of your head. That's another $319.4 billion.

For $44.3 billion on amusement/theme parks, see Global Industry Analysts, "The Global Theme/Amusement Parks Market: Trends, Drivers & Projections," November 2015. For $12.1 billion in retail sales of action sports products in the United States, see Active Marketing Group, "The Action Sports Market," 2007. For $263 billion spent on adventure travel, see Adventure Travel Trade Association, "Adventure Travel: A Growing, but Untapped Market for Agents," February 17, 2015, http://www.travelmarketreport.com/articles/Adventure-Travel-A-Growing-But-Untapped-Market-for-Agents. For meditation: Jan Wieczner, "Meditation Has Become a Billion-Dollar Business," *Fortune*, March 12, 2016. For yoga: "2014 Outlook for the Pilates and Yoga Industry," *SNews*, December 16, 2013, http://www.snewsnet.com/news/2014-yoga-pilates-studios/.

Finally, we added in casino gambling. For sure, we could make the argument that much of this industry is motivated by the heavy dopamine spikes produces by gambling, but we needed to discount sports betting and a few other categories where there are other reasons (profit) for betting. But everything about a casino—from the lack of clocks to the pumped-in oxygen—is designed to get us out of our heads and immersed in the game. This put in an additional $159.71 billion. See Statistica, "Statistics and Facts About the Casino Industry," 2015, https://www.statista.com/topics/1053/casinos/.

Thus the current grand total is $3.99 trillion.

32 Galaxies in the entire universe: Henry Fountain, "Two Trillion Galaxies, at the Very Least," *New York Times*, Oct. 17, 2016.

Chapter Two: Why It Matters

33 In 2011, an out-of-work television host named Jason Silva: All the details of Jason Silva's life that appear in this chapter come from a series of 'author interviews conducted between 2014 and 2016.

33 Titled "You Are a Receiver": https://vimeo.com/27668695.

33 The *Atlantic* ran a long profile: Ross Anderson, "A Timothy Leary for the Viral Video Age," *Atlantic*, April 12, 2012.

35 Al Gore's network: For a look at Silva on Current, check out https://vimeo.com/6950613.

36 Certainly, researchers have come up with plenty of other descriptions of altered states, but we chose the four categories of STER for a specific reason: There are a lot of different breakdowns to consider, but in coming up with our analysis, the following were key: Charles Tart, *Altered States of Consciousness* (New York: Harper, 1990); William James, *Varieties of Religious Experience* (N.p.: CreateSpace, 2009); Mihaly Csikszentmihalyi, *Flow* (New York: Harper Perennial, 2008), pp. 43–93;

Bruce Greyson, "The Near-Death Experience Scale," *Journal of Nervous and Mental Disease* 171, no. 6 (1983): 369–74; Erich Studerus, Alex Gamma, and Franz Vollenweider, "Psychometric Evaluation of the Altered States of Consciousness," *PLoS One*, August 31, 2010; Ronald Pekala, *Quantifying Consciousness* (Boston: Springer, 1991). Also see Rick Strassman, "Hallucinogenic Drugs in Psychiatric Research and Treatment," *Journal of Nervous and Mental Disease* 183, no. 3 (1995): 127–38, Robert Thurman, *The Tibetan Book of the Dead: The Great Book of Natural Liberation Through Understanding in the Between* (New York: Bantam Books, 1993).

37 "[T]he self is not an unmitigated blessing": Mark Leary, *The Curse of the Self* (Oxford: Oxford University Press, 2007), p. 21.

38 Scientists call this shutdown: Arne Dietrich, "Functional Neuroanatomy of Altered States of Consciousness," *Conscious Cognition,* June 12, 2003, pp. 231–56.

38 Psychologist Bob Kegan: See both Robert Kegan, *The Evolving Self* (Cambridge, MA: Harvard University Press, 1982), and Robert Kegan, *In Over Our Heads* (Cambridge, MA: Harvard University Press, 1994).

39 According to a 2015 Gallup survey: Frank Newport, "Americans' Perceived Time Crunch No Worse Than Past," Gallup.com, December 31, 2015, http://www.gallup.com/poll/187982/americans-perceived-time-crunch-no-worse-past.aspx.

40 "Time poverty": Maria Konnikova, No Money, No Time," *New York Times*, June 13, 2014.

40 Our sense of time isn't localized: Much of the timelessness information comes from an author interview with David Eagelman, 2012. Also see David Eagleman, *Incognito* (New York: Pantheon Books, 2011), pp. 51–54.

40 In his book *The Time Paradox:* Philip Zimbardo, *The Time Paradox* (New York: Atria, 2009), p. 261.

40 In a recent study published in *Psychological Science*: Melanie Rudd, Kathleen Vohs, and Jennifer Aaker, "Awe Expands People's Perception of Time, Alters Decision Making, and Enhances Well-Being," *Psychological Science* 23, no. 10 (2011): 1130–36.

41 "[M]emory distortions are basic": Elizabeth Loftus et al., "False Memories in Highly Superior Autobiographical Memory Individuals," *PNAS* 110, no. 52 (2013): 20947–52.

42 One in three Americans, for example, is obese: National Institutes of Diabetes and Digestive and Kidney Diseases, https://www.niddk.nih.gov/health-information/health-statistics/Pages/overweight-obesity-statistics.aspx.

42 Big-box health clubs oversell memberships by 400 percent: Confidential author interview with senior vice president of Lifetime Fitness, June 2015.

42 And when a Harvard Medical School study confronted patients: Robert Kegan, *Immunity to Change: How to Overcome It and Unlock the Potential in Yourself and Your Organization* (Boston: Harvard Business Press, 2009), p. 1.

42 In flow, as in most of the altered states: Steven Kotler, *The Rise of Superman* (Boston: New Harvest, 2013). Also see www.flowgenomeproject.com/stealingfiretools".

42 "The [experience] lifts the course of life to another level": Mihaly Csikszentmihalyi, *Flow* (New York: Harper Perennial, 2008), p. 69.

43 So many people find this so great and high: Abraham Maslow, *Religion, Values, and Peak Experiences* (New York: Penguin, 1994), p. 62.

43 "In a culture supposedly ruled by the pursuit": Mihaly Csikszentmihalyi, *Beyond Boredom and Anxiety* (San Francisco: Jossey-Bass, 2000), p. 197.

43 When John Hagel: Author interview with John Hagel, 2014.

44 In his first video, "You Are a Receiver": https://vimeo.com/27668695.

44 William James experienced this during his Harvard experiments: William James, *The Varieties of Religious Experience* (New York: Create Space, 2009), p. 374.

44 Often, an ecstatic experience: This overview is mainly intended to track what happens in the brain as we enter a flow state. In other instances of ecstasis, like meditation, you can get a decrease in cortico-adrenal response, a boost in DHEA and GABA and other neurochemicals in a different sequence than we are describing here. That said, as the below notes make clear, there is considerable overlap with what happens in flow as in meditation and mystical states and psychedelic states. For a solid layperson's overview of how flow changes the brain, see Herbert Benson and William Proctor, *The Breakout Principle* (New York: Scribner, 2003), pp. 46–58. For a little more tech-

nical approach, see Dieter Vaitl et al., "Psychobiology of Altered States of Consciousness," *Psychological Bulletin* 131, no. 1 (2005): 98–127, and M. Bujatti, "Serotonin, Noradrenaline, Dopamine Metabolites in Transcendental Meditation-Technique," Journal of Neural Transmission 39, no 3 (September 1976): 257–67. It should also be emphasized this is a broad overview, with lots of questions remaining to be solved as measurement tools improve and researchers start to conduct comparative studies between different ecstatic techniques.

44 These neurochemicals raise heart rates:" Raja Parasuraman, *The Attentive Brain* (Cambridge, MA: A Bradford Book/MIT Press, 2000), pp. 34–44. A good general-purpose overview of both norepinephrine and dopamine can be found in Helen Fisher, *Why We Love: The Nature and Chemistry of Romantic Love* (New York: Henry Holt, 2004). For a great discussion of dopamine's role in flow, see Gregory Burns, *Satisfaction: The Science of Finding True Fulfillment* (New York: Henry Holt, 2005), pp. 146–74.

44 These chemicals amp up the brain's pattern recognition abilities: P. Krummenacher, C. Mohr, H. Haker, and P. Brugger, "Dopamine, Paranormal Belief, and the Detection of Meaningful Stimuli," *Journal of Cognitive Neuroscience* 22, no. 8 (2010): 1670–81; Georg Winterer and Donald Weinberger, "Genes, Dopamine, and Cortical Signal-to-Noise Ration in Schizophrenia," *Trends in Neuroscience* 27, no. 11 (November 2004); and S. Kroener, L. J. Chandler, P. Phillips, and Jeremy Seamans, "Dopamine Modulates Persistent Synaptic Activity and Enhances the Signal to Noise Ratio in the Prefrontal Cortex," *PLoS One* 4, no. 8 (August): e6507. Also see Michael Sherman's great talk on how dopamine/pattern recognition lead to strange beliefs: http://www.ted.com/talks/michael_shermer_on_believing_strange_things?language=en.

44 As these changes . . . are taking place: The literature surrounding altered states of consciousness and brainwaves is considerable. For a general review of brainwave function, see Ned Harrrmann, "What Is The Function of Various Brainwaves," ScientificAmerican.com, December 22, 1997, https://www.scientificamerican.com/article/what-is-the-function-of-t-1997-12-22/.

For a good overview of brainwave activity during flow, see

Sally Adee, "Zapping the Brain to Get into Flow," *Washington Post*, February 13, 2012. Also Steven Kotler, *The Rise of Superman* (Boston: New Harvest, 2013), pp. 32–41. Lastly, a paper that isn't yet peer-reviewed but is interesting in its findings: Jan Van Looy et al., "Being in the Zone: Using Behavioral and EEG Recording for the Indirect Assessment of Flow," available at https://peerj.com/preprints/2482.pdf. For psychedelics: E. Hoffmann, J. M. Keppel Hesselink, and Yatra-W.M. da Silveria Barbosa, "Effects of a Psychedelics, Tropical Tea, Ayahuasca, on the EEG Activity of the Human Brain during a Shamanic Ritual," *MAPS Magazine*, Spring 2001. For contemplative and spiritual practices: Yuji Wada et al., "Changes in EEG and Autonomic Nervous Activity During Meditation and Their Association with Personality Traits," *International Journal of Psychophysiology* 55, no. 2 (February 2005): 199–207. Also, J. Polichj and B. R. Cahn, "Meditation States and Traits," *Psychological Bulletin* 132, no. 2 (2006): 180–211.

45 Then parts of the prefrontal cortex begin shutting down: For psychedelics: R. L. Carhart-Harris et al., "Neural Correlates of the Psychedelic State as Determined by fMRI Studies with Psilocybin," PNAS, 2012, doi/10.1073/pnas.1119598109. For flow: Arne Dietrich, "Functional Neuroanatomy of Altered States of Consciousness," *Conscious Cognition,* June 12, 2003, pp. 231–56. For meditation: Judson Brewer et al., "Meditation Experience Is Associated with Difference in Default Mode Network Activity and Connectivity," *PNAS* 108, no. 50 (2011): 20254–59. Judson Brewer also gave a good TEDx talk on his work, available at https://www.youtube.com/watch?v=jE1j5Om7g0U. For a thorough review of the neurobiology of out-of-body experiences and other "mystical" phenomena, see Andra Smith and Claude Messier, "Voluntary Out-of-Body Experience: An fMRI Study, *Frontiers of Human Neuroscience,* February 10, 2014. And for a great book on the early research into the neuroanatomical changes produced by spiritual experiences: Andrew Newberg and Eugene D'Aquilli, *Why God Won't Go Away* (New York: Ballantine Books, 2002).

45 . . . brain releases endorphins and anandamide: The literature on neurochemistry and altered states is considerable, but for a good overview see both J. Allan Hobson, *The Dream Drugstore* (Cambridge, MA: MIT Press, 2001), and Dean Hamer,

The God Gene (New York: Anchor, 2005). For endorphins, see James Henry, "Possible Involvement of Endorphins in Altered States of Consciousness," *Ethos* 10 (1982): 394–408; Henning Boecker et al., "The Runner's High: Opiodergic Mechanisms in the Human Brain," *Cerebral Cortex* 18, no. 11 (2008): 2523–31; A. Dietrich and W. F. McDaniel, "Endocannabinoids and Exercise," *British Journal of Sports Medicine* 38 (2004): 536–44. Also see Etzel Cardena and Michael Winkelman, *Altering Consciousness* (Santa Barbara, CA: Praeger, 2011), p. 171. A good overview of different psychedelic plants and their impact on neurochemistry and spiritual experience can be found in Elaine Perry and Valerie Laws's article, "Plant Gods and Shamanic Journeys," in *New Horizons in the Neuroscience of Consciousness* (Philadelphia: John Benjamins, 2010), pp. 309–24. Lastly, Boecker, "Brain Imaging Explores the Myth of Runner's High."

45 Anandamide also plays another important role here: S. Hao, Y. Avraham, R. Mechoulam, and E. Barry, "Low Dose Anandamide Affects Food Intake, Cognitive Function, Neurotransmitter and Corticosterone Levels in Diet-Restricted Mice," *European Journal of Pharmacology* 392, no. 3 (March 31, 2000): 147–56.

45 . . . an afterglow of serotonin and oxytocin: Lars Farde et al., "The Serotonin System and Spiritual Experiences," *American Journal of Psychiatry* 160 (2003): 1965–69; Umit Sayin, "Altered States of Consciousness Occurring During Expanded Sexual Response in the Human Female," *NeuroQuantology* 9, no. 4 (2011); N. Goodman, "The Serotonergic System and Mysticism," *Journal of Psychoactive Drugs* 34, no. 3 (2002): 263–72. Also, John Ratey and Eric Hagerman, *Spark: The Revolutionary New Science of Exercise and the Brain* (New York: Little, Brown, 2008).

45 Conscious processing can handle only about 120: Mihaly Csikszentmihalyi, *Flow* (New York: Harper Perennial, 2008), pp. 28–30. Also, for one of the best books on this subject, see Tor Norretranders, *The User Illusion: Cutting Consciousness Down to Size* (New York: Penguin, 1990).

45 *Umwelt* is the technical term: Stanford neuroscientist David Eagleman gives a great, detailed explanation of the neuroscience of our umwelt in "Can We Create New Senses," his 2015 TED talk: https://www.ted.com/talks/david_eagleman_can_we_create_new_senses_for_humans?language=en.

46 Red Bull Hacking Creativity Project: The project has a great website where many of their findings are described: http://hackingcreativity.com.

47 "wicked problems" of today: Jon Kolko, "Wicked Problems: Problems Worth Solving, *Stanford Social Innovation Review*, March 2012. Also, John Camillus, "Strategy as a Wicked Problem," *Harvard Business Review*, May 2008.

47 The ability to face constructively the tension: Roger Martin, *The Opposable Mind: Winning Through Integrative Thinking* (Boston: Harvard Business School Press, 2009), p. 15.

47 Research done on Tibetan Buddhists: Kathy Gilsinan, "The Buddhist and the Neuroscientist," *Atlantic*, July 4, 2015.

47 They arise primarily during binding: John Kounios and Mark Beeman, "The Cognitive Neuroscience of Insight," *Annual Review of Psychology* 65 (2009): 71–93.

48 Initial studies showed eight weeks of meditation: Fadel Zeidan, Susan Johnson, Bruce Diamond, and Paula Goolkasian, "Mindfulness Meditation Improves Cognition: Evidence of Brief Mental Training," *Consciousness and Cognition* (2010); Lorenza S. Colzato, Ayca Szapora, Dominique Lippelt, and Bernhard Hommel, "Prior Meditation Practice Modulates Performance and Strategy Use in Convergent- and Divergent-Thinking Problems," *Mindfulness* (2014) doi:10.1007/s12671-014-0352-9.

48 "Simply stated": "Brief Meditative Exercise Helps Cognition," *Science Daily*, April 19. 2010, https://www.sciencedaily.com/releases/2010/04/100414184220.htm.

48 In a recent University of Sydney study: Richard Chi and Allan Snyder, "Brain Stimulation Enables Solution to Inherently Difficult Problem," *Neuroscience Letters* 515 (2012): 121–24.

49 When neuroscientists at DARPA and Advanced Brain Monitoring: ABM CEO Chris Berka gave a great TEDx talk about this research: http://tedxtalks.ted.com/video/Whats-next-a-window-on-the-brai;TEDxSanDiego. Also, "9-Volt Nirvana," Radiolab, June 2014, http://www.radiolab.org/story/9-volt-nirvana/; Sally Adee, "Zap Your Brain into the Zone," *New Scientist*, February 1, 2012.

49 Several decades ago, James Fadiman: James Fadiman, *The Psychedelic Explorer's Guide: Safe, Therapeutic, and Sacred Journeys* (Rochester, VT: Park Street Press, 2011), p. 133.

50 Author and venture capitalist Tim Ferriss: Author interview with Tim Ferriss, 2016.

50 As Ferriss explained on CNN: "Can LSD Make You a Billionaire," CNN, January 25, 2015. See https://www.youtube.com/watch?v=jz9yZFtRJjk.

50 A 500 percent boost in productivity: Susie Cranston and Scott Keller, "Increasing the 'Meaning Quotient' of Work," *McKinsey Quarterly,* January 2013.

Chapter Three: Why We Missed It

51 The English Pale: There are lots of potential references, but in this case Wikipedia does a really solid job: https://en.wikipedia.org/wiki/The_Pale.

52 James Valentine is a tall, thin man: Author interviews with James Valentine, 2015 and 2016.

54 Consider Joan of Arc: If you'd like to consider Joan of Arc try Mark Twain, *Personal Recollections of Joan of Arc* (San Francisco: Ignatius Press, 1989), and George Bernard Shaw, *Saint Joan* (New York: Penguin, 2001).

55 At her trial for heresy: Regine Pernoud, "*Joan of Arc by Herself and Her Witnesses,* trans. Edward Hyams (New York: Stein & Day, 1962)," p. 184.

56 The prejudice that whatever matters: Andy Clark, *Natural-Born Cyborgs: Minds, Technologies, and the Future of Intelligence* (New York: Oxford University Press, 2004), p. 5

56 In 2012, a study conducted: American Academy of Pediatrics, May 1, 2014. See EurekaAlert: https://www.eurekalert.org/pub_releases/2014-05/aaop-mil042514.php.

56 By 2015, that number had jumped to one in three: Nita Farahany, "Colleges Should Allow Students to Take Smart Drugs," *Washington Post*, November 3, 2015. See also, "Monitoring the Future 2013 Survey Results: College and Adults," National Institute of Drug Abuse, https://www.drugabuse.gov/related-topics/trends-statistics/infographics/monitoring-future-2013-survey-results-college-adults.

57 "Are Smart Drugs Cheating?": Karen Weintraub, "Some Students Don't See ADHD Drug Use as Cheating," *USA Today,* May 3, 2014; Princess Ojiaku, "'Smart Drugs' Are Here—Should College Students Be Allowed to Use Them?,"

Washington Post, November 3, 2015; "Is Taking Smart Drugs Cheating?," *Newsy*, November 17, 2015; Matt Tinoco, "Are You a Cheater If You're Using Smart Drugs to Get Ahead," *Vice*, April 13, 2005.

57 "archaic techniques of ecstasy": Mircea Eliade, *Shamanism: Archaic Techniques of Ecstasy*, Bollingen Series (Princeton, NJ: Princeton University Press, 2004).

57 "Narcotics, he argued": Ibid., p. 401.

57 "The ultimate wisdom of enlightenment": Sam Harris, *Waking Up* (New York: Simon & Schuster, 2015), p. 124.

58 Walter Pahnke conducted: For a good review of the experiment, see John Horgan, *Rational Mysticism* (Boston: Mariner Books, 2004).

58 "[Psilocybin] subjects ranked their experiences": Ibid. p. 27.

59 Researchers have twice gone back to double-check Pahnke's: Rick Doblin, "Pahnke's 'Good Friday Experiment': A Long Term Follow-Up and Methodological Critique," *Journal of Transpersonal Psychology* 23, no. 1 (1991); Roland Griffiths et al., "Psilocybin Can Occasion Mystical-Type Experiences Having Substantial and Sustained Personal Meaning and Spiritual Significance," *Psychopharmacology* 187 (2006), doi:10.1007/s00213-006-0457-5.

59 When author Michael Pollan asked him: Michael Pollan, "The Trip Treatment," *New Yorker*, February 9, 2015.

59 neuroscientist Michael Persinger's God Helmet: Jack Hitt, "This Is Your Brain on God," *Wired*, November 1, 1999.

59 Palo Alto Neuroscience: Author interview with Mikey Siegel (who advises the company), 2015. Also see www.paloaltoneuroscience.com.

60 David Nutt's: For the entire equasy and ecstasy story see David Nutt, *Drugs—Without the Hot Air* (Cambridge, England: UIT Cambridge, 2012), pp. 1–30.

61 Headlines across the country read: For example, Christopher Hope, "Ecstasy 'No More Dangerous than Horse Riding,'" *Telegraph*, February 7, 2009.

61 Ecstasy is a harmful drug: Nutt, *Drugs—Without the Hot Air*, p. 20.

62 exchange between Nutt and the home secretary: Ibid, pp. 20–21.

63 "Government minister claims alcohol more dangerous than LSD!": Jack Doyle, "Alcohol More Dangerous than LSD, Says Drug Advisor," *Independent*, October 29, 2009.

64 Aside from some foul cutting material: Alexander Zaitchik, The Speed of Hypocrisy," *Vice,* June 30, 2014.

64 The 1.2 million Americans who tried meth: This is from the National Institutes of Drug Abuse 2012 data, https://www.drugabuse.gov/publications/research-reports/methamphetamine/letter-director.

64 Intel researcher and author Melissa Gregg: Melissa Gregg, "The Neverending Workday," *Atlantic*, October 15, 2015.

65 Hamelin is a town of about fifty thousand people: The town of Hamelin has a great website detailing the Pied Piper story: http://www.hameln.com/tourism/piedpiper/rf_sage_gb.htm.

65 According to the Luneburg Manuscript: "The Disturbing True Story of the Pied Piper of Hamlin," Ancient Origins, August 14, 2014, http://www.ancient-origins.net/myths-legends/disturbing-true-story-pied-piper-hamelin-001969?nopaging=1.

66 Historians continue to debate the Pied Piper of Hamelin: See, for example, http://www.medievalists.net/2014/12/07/pied-piper-hamelin-medieval-mass-abduction/.

67 "[It's] a sense of total annihilation": Stanislav Grof, *The Adventure of Self-Discovery* (Albany: State University of New York Press, 1988), p. 30.

68 Why evangelical megachurches are booming (with over six million attendees every Sunday): Jesse Bogan, "America's Biggest Megachurches," *Forbes*, June 26, 2009; C. Kirk Hadaway and P. L. Marler, "Did You Really Go to Church This Week: Behind the Poll Data," *Christian Century,* May 6, 1998.

68 *Communitas*" is the term: Victor Turner, *The Ritual Process: Structure and Anti-Structure* (Chicago: Aldine, 1969), pp. 94–113

68 "Exaggerations of communitas": Ibid., p. 129.

69 "I am beginning to comprehend": "The Triumph of Hitler," History Place, http://www.historyplace.com/worldwar2/triumph/tr-will.htm.

69 According to Fuhrer confidant, Ernst Hanfstaengl": Peter Conradi, *Hitler's Piano Player: The Rise and Fall of Ernst Hansfstaengl, Confidant of Hitler, Ally of FDR*" (New York: Carroll & Graff, 2004), p. 45.

Chapter Four: Psychology

73 As Archbishop of Canterbury John Tillotson later noted": John Tillotson, *Sermons Preached on a Number of Occasions* (London: For S. A. Gellibrand, 1674).

75 Oprah Winfrey teamed up": Jesse McKinley, "The Wisdom of the Ages, for Now Anyway," New York Times, March 23, 2008.

75 A New Earth attracted ten million: Oliver Burkeman, "The Bedsit Epiphany," *Guardian*, April 10, 2009; Jennifer Fermino, "Pope Francis Tells Crowd of 20,000 'God Is Living in Our Cities' at Madison Square Garden Mass, Closing Out NYC Visit," New York *Daily News*, September 25, 2005; Alice Philipson, "The Ten Largest Gatherings in History," *Telegraph*, January 19, 2015.

75 He grew up in the rubble of postwar: "Eckhart Tolle Biography," *New York Times*, Times Topics, March 5, 2008; Eckhart Tolle, *The Power of Now* (Novato, CA: Namaste, 2004).

77 This was the era of *The Man in the Gray Flannel Suit*": Sloan Wilson, *The Man in the Gray Flannel Suit* (Boston: Da Capo Press, 2002).

77 "The Beat Generation was a vision that we had": Jack Kerouac, "Aftermath: The Philosophy of the Beat Generation," *Esquire,* March 1958.

77 Allen Ginsberg's epic poem: Allen Ginsberg, *Howl and Other Poems* (San Francisco: City Lights, 2001).

77 Harvard of the Human Potential Movement: Gregory Dicum, "Big Sur without the Crowds," *New York Times,* January 7, 2007.

77 So central was Esalen: Emily Nussbaum, "The Original Existentially Brilliant 'Mad Men' Finale"," *New Yorker,* May 18, 2015.

78 largely laid by Dick Price: Jeffrey Kripal, *Esalen: America and the Religion of No Religion* (Chicago: University of Chicago Press, 2008), p. 69.

79 notes author and modern religious historian Erik Davis: Lee Gilmore, Mark Van Proyen, *AfterBurn: Reflections on Burning Man* (Albuquerque: University of New Mexico Press, 2005), p. 23.

79 As an early Esalen motto put it: Kripal, *Esalen*, p. 9.

79 So Erhard repackaged an assortment of Esalen: Erhard remains a polarizing figure and his career is subject to competing and conflicting interpretations. For a couple on either side of that debate, see Steven Pressman, Outrageous Betrayal: The Dark Journey of Werner Erhard from est to Exile (New York: St. Martin's Press, 1993), and William Bartley, *Werner Erhard:*

The Transformation of a Man (New York: Clarkson Potter, 1988).

80 boasts corporate clients including Microsoft: Peter Haldemannov, "The Return of Werner Erhard, Father of Self-Help," *New York Times*, November 28, 2015.

82 they describe their central practice as OMing: www.onetaste.us.

82 "The search for personal transformation": Patricia Leigh Brown and Carol Pogash, "The Pleasure Principle," *New York Times*, March 13, 2009.

82 To put this in perspective, Planned Parenthood founder Margaret Sanger: "Women Working 1800–1930," Harvard University Library Open Collections Program, http://ocp.hul.harvard.edu/ww/sanger.html. And it didn't stop there. Morality, especially of the sexual stripe, has been actively legislated. In Mississippi, for example, you are still forbidden to *explain* polygamy, let alone practice it. In Arizona, they prudently limit the number of sex toys to two per house.

83 University of Pennsylvania neurologist: Anjan Chatterjee, *The Aesthetic Brain: How We Evolved to Desire Beauty and Enjoy Art* (New York: Oxford University Press, 2013), p. 88.

83 Social scientist Jenny Wade: Jenny Wade, *Transcendent Sex: When Lovemaking Opens the Veil* (New York: Gallery Books; 2004).

84 the French research group Sexualitics: Antoine Mazieres, Mathieu Trachman, Jean-Philippe Cointet, Baptiste Coulmont, and Christophe Prieur, "Deep Tags: Toward a Quantitative Analysis of Online Pornography," *Porn Studies* 1, nos. 1–2 (2014): 80–95.

84 But in 2010: Meriss Nathan Gerson, "BDSM vs. the DSM," *Atlantic Monthly*, January 13, 2015.

85 one of the fastest-selling books in history: Anita Singh, "50 Shades of Grey is Bestselling Book of All Time," *Telegraph*, August 7, 2012.

85 Pressure in the throat or colon regulates the vagus nerve: Anish Sheth and Josh Richman, *What's Your Poo Telling You?* (San Francisco: Chronicle Books, 2007).

85 A 2013 Dutch study found that kinky sex: A. A. J. Wismeijer and M. A. L. M. Van Assen, "Psychological Characteristics of BDSM Practitioners," *Journal of Sexual Medicine* (August 2013): 1943–52.

86 Megachurch minister Ed Young: Gretel C. Kovachnov, "Pastor's Advice for Better Marriage: More Sex," *New York Times*, November 23, 2008.

87 In the 1990's Britton became interested in near-death experiences: Willoughby Britton and Richard Bootzin, "Near-Death Experiences and the Temporal Lobe," *Journal of Psychological Science* 15, no. 4 (April 2004): 254–58. Also, the original author interview with Britton took place for Steven Kotler, "Extreme States," *Discover*, July 2005, http://discovermagazine.com/2005/jul/extreme-states.

87 Then Johns Hopkins neuroscientist Roland Griffiith: Katherine A. MacLean, Matthew W. Johnson, and Roland R. Griffiths, "Mystical Experiences Occasioned by the Hallucinogen Psilocybin Lead to Increases in the Personality Domain of Openness," *Journal of Psychopharmacology* 25, no. 11 (November 2011): 1453–61; R. Griffiths, W. Richards, U. McCann, and R. Jesse, "Psilocybin Can Occasion Mystical Experiences Having Substantial and Sustained Personal Meaning and Spiritual Significance," *Psychopharmacology* 187 (2006): 268–83. Also see Lauren Slater, "How Psychedelic Drugs Can Help Patients Face Death," *New York Times*, April 20, 2012; Steven Kotler, "The Psychedelic Renaissance," *Playboy*, April 2011.

88 In 2012, psychologist Michael Mithoefer discovered: Michael C Mithoefer, Mark T Wagner, Ann T Mithoefer, Lisa Jerome, and Rick Doblin, "The Safety and Efficacy of ±3,4-methylenedioxymethamphetamine-Assisted Psychotherapy in Subjects with Chronic, Treatment-Resistant Posttraumatic Stress Disorder: The First Randomized Controlled Pilot Study," *Journal of Psychopharmacology* 25, no. 4 (April 2011): 439–52.

90 In her book *Train Your Mind, Change Your Brain*: Sharon Begley, *Train Your Mind, Change Your Brain* (New York: Ballantine Books, 2007), p. 250.

90 University of Pennsylvania psychologist Martin Seligman: Rob Hirtz, "Martin Seligman's Journey from Learned Helplessness to Learned Happiness," *Pennsylvania Gazette*, January/February 1999.

91 In 1998, after being elected: Martin Seligman, "Building Human Strength: Psychology's Forgotten Mission," *American Psychological Association Newsletter* 29, no. 1 (January 1998).

92 Kegan spent three decades tracking this group: Robert Ke-

gan, *In Over Our Heads* (Cambridge, MA: Harvard University Press, 1994). Kegan's work is vast, dense, and worthy of extended study for anyone interested in the field of adult development. Along with coauthor Lisa Lahey and others, he has also written *Immunity to Change* and *How the Way We Talk Can Change the Way We Work* (among countless other academic papers and monographs). We have deliberately streamlined his categories of higher adult development for the lay reader, combining transitional self-authoring and self-transforming, or what his colleague Bill Torbert calls the strategist and alchemist stages into that leading 5 percent that we reference. For additional models of adult development, see Harvard Business School's Ron Heifetz's notion of technical versus adaptive leadership, Chris Argyris, Susanne Cook-Greuter, Bill Torbert, et al. For a current dive into the scholarship and debates within the field, Angela H. Pfaffenberger, ed., *The Postconventional Personality* (Albany: State University of New York Press, 2011) is a good place to start.

92 But in all of this developmental research: Author interview with Susanne Cook-Greuter, March 2009. This was literally in the footnotes of one of the earliest studies of advanced stages of adult development—the ".01%ers" who had reportedly tested as having stabilized postconventional consciousness. As the sample size was tiny—less than ten people—these reports can only be taken anecdotally, but they are nonetheless intriguing. See also Paul Marko, "Exploring Facilitative Agents That Allow Ego Development to Occur," in Pfaffenberger, ed., *The Postconventional Personality*, p. 99. For another survey of the role of peak states and advanced development, see Allan Badiner and Alex Grey, *Zig Zag Zen: Buddhism and Psychedelics* (San Francisco: Chronicle Books, 2002).

92 Fifty years ago, psychologist Abraham Maslow: Abraham Maslow, *Toward a Psychology of Being* (Princeton, NJ: Princeton University Press, 1962).

93 A 2012 study published in *Cognitive Processing* took it further: Referenced in Pfaffenberger, ed. *The Postconventional Personality*, p. 27. Primary citations: S. Harung, F. Travis, A. M. Pensgaard, R. Boes, S. Cook-Greuter, and K. Daley, "Higher Psycho-Physiological Refinement in World-Class Norwegian Athletes: Brain Measures of Performance Capacity,"

Scandinavian Journal of Medicine and Science in Sports 21, no. 1 (February 2011): 32–41. See also Susanne Cook-Greuter, "Making the Case for a Developmental Perspective," Industrial and Commercial Training 36, no. 7 (2004): 275–81; H. S. Harung and F. Travis, "Higher Mind-Brain Development in Successful Leaders: Testing a Unified Theory of Performance," *Journal of Cognitive Processing* 13 (2012): 171–81. See also Robert Panzarella, "The Phenomenology of Aesthetic Peak Experiences" (Ph.D. diss., City University of New York, 1977), for an additional study involving musicians and artists that demonstrated a relationship between frequency of peak states and overall "self-actualization."

93 Boston College's Bill Torbert found: David Rooke and William Torbert, "Seven Transformations of Leadership," *Harvard Business Review*, April 2005, p. 7.

Chapter Five: Neurobiology

95 There's a bit of southern folk wisdom: Raff Viton and Michael Maddock, "Innovating Outside the Jar," *Bloomberg*, July 29, 2008.

96 Nicole Kidman: Julia Neel, "The Oscars 2011," *Vogue*, March 1, 2011.

96 When seriously depressed patients received Botox injections: Lenny Bernstein, "Using Botox to Treat Depression. Seriously," *Washington Post*, May 7, 2014.

96 But when Botoxed subjects were asked to empathize: Pamela Paul, "With Botox, Looking Good and Feeling Less," *New York Times*, June 17, 2011.

97 Our facial expressions are hardwired: Paul Ekman, *Emotions Revealed* (New York: Holt, 2007), pp. 1–16.

97 The body, the gut, the senses: Jack Meserve, "Your Brain and Your Body Are One and the Same," *New York*, November 19, 2015.

97 we're smart *because* we have bodies: For one of the best breakdowns of embodied cognition, see Guy Claxton, *Intelligence in the Flesh* (New Haven, CT: Yale University Press, 2015). Also see Samuel McNerney, "A Brief Guide to Embodied Cognition," *Scientific American*, November 4, 2011.

98 If someone gave you a cup of icy cold water: Lawrence Williams and John Bargh, "Experiencing Physical Warmth Pro-

motes Interpersonal Warmth," *Science*, October 24, 2008, pp. 606–7.

98 Or consider Harvard psychologist Amy: https://www.ted.com/talks/amy_cuddy_your_body_language_shapes_who_you_are?language=enhot and Amy Cuddy, *Presence: Bringing Your Boldest Self to Your Biggest Challenges* (New York: Little, Brown, 2015).

99 Weightlessness, weightedness, and rotation: Author interview with Jimmy Chin, 2013.

99 "[movement sequences] have an impact on stress": Peter Strick et al., "Motor, Cognitive, and Affective Areas of Cerebral Cortex Influence the Adrenal Medulla," *PNAS* 113, no. 35 (2016): 9922–27

100 Her name is Ellie: A demonstration of Ellie at work is available at https://www.youtube.com/watch?v=ejczMs6b1Q4.

101 constant state of impression management: Author interview with Skip Rizzo, 2015.

102 And people prefer talking to Ellie: Tanya Abrams, "Virtual Humans Inspire Patients to Open Up, USC Study Suggests," *USC News*, July 9, 2014.

103 In 1999 Steven Spielberg had a problem: "Inside *Minority Report*'s Idea Summit, Visionaries Saw The Future," Wired.com, June 6, 2012, https://www.wired.com/2012/06/minority-report-idea-summit/.

104 Chris Berka, the founder of Advanced Brain Monitoring: Author interview with Chris Berka, 2015.

105 In a related study run in Barcelona: Author interview with Chris Berka, 2015. ESADE has yet to publish these results, but Thomas Maak, ESADE professor of people management and organization, gave a 2013 TEDx talk about the study: https://www.youtube.com/watch?v=CvOLbYChYcw&spfreload=5.

106 Dr. Andrew Newberg doesn't look like a rebel: Author interviews with Andrew Newberg, 2015–16. Also see Steven Kotler, "The Neurology of Spiritual Experience," *HPlus*, September 16, 2009. If you're really curious about early research in neurotheology and looking for a less technical introduction, check out Andrew Newberg and Mark Waldman, *How God Changes the Brain* (New York: Ballantine Books, 2010). More technical is Andrew Newberg, *Principles of Neurotheology* (Burlington, VT: Ashgate, 2010).

109 Arzy, in 2011, became interested: Author interview with Shahar Arzy, 2016. See also Shahar Arzy and Moshe Idel, *Kabbalah: A Neurocognitive Approach to Mystical Experiences* (New Haven, CT: Yale University Press, 2015). Also, for a more general work on the brain's ability to produce autoscopic phenomena (out-of-body experiences, doppelganger experiences), see Anil Ananthaswamy, *The Man Who Wasn't There: Investigations into the Strange New Science of the Self* (New York: Dutton, 2015).

109 The Jewish mystic Abraham Abulafia: Abulafia is one of the most interesting figures in Jewish mysticism. A lot has been written, but a great place to start is Moshe Idel, *The Mystical Experience in Abraham Abulafia* (Albany: State University of New York Press, 1987).

109 He also found a way to study this phenomenon in regular subjects: This experiment was really a variation on earlier work done by Olaf Blanke (Arzy's Ph.D. thesis advisor). See Olaf Blanke, Bigna Lenggenhager, and Jane Aspel, "Keeping in Touch with One's Self: Multisensory Mechanisms of Self Consciousness," *PloS One,* August 5, 2009, http://dx.doi.org/10.1371/journal.pone.0006488.

111 Abraham Maslow once famously said: Abraham Maslow, *The Psychology of Science: A Reconnaissance* (Chapel Hill, NC: Maurice Bassett, 2004), p. 15

111 the one in four Americans now on psychiatric medicines: "Total Number of People Taking Psychiatric Drugs in the United States," *CCHR International*, https://www.cchrint.org/psychiatric-drugs/people-taking-psychiatric-drugs/.

111 the escalating rate of suicide: Sally Curtin, Margaret Warner, and Holly Hedegaard, "Increases in Suicide in the United States, 1999–2014," NCHS Data Brief No. 241, April 2016, http://www.cdc.gov/nchs/products/databriefs/db241.htm.

112 We could get on a treadmill: The data on this one is pretty overwhelming. For a solid overview see John Ratey and Eric Hagerman, *Spark: The Revolutionary New Science of Exercise and the Brain* (New York: Little, Brown, 2013).

112 get some natural sunshine: Michael Holick, "Vitamin D Deficiency," *New England Journal of Medicine* 357 (2007): 266–81.

112 practice meditation for fifteen minutes: Jennifer Haythornthwaite et al., "Meditation Programs for Psychological Stress and

Well-being: A Systematic Review and Meta-analysis," *JAMA Internal Medicine* 174, no. 3 (2014): 357–68.

113 Tibetan monks can shut off: Judson Brewer et al., "Meditation Experience Is Associated with Difference in Default Mode Network Activity and Connectivity," *PNAS* 108, no. 50 (December 13, 2011): 20254–59.

113 SEAL snipers tune their brainwaves to the alpha frequency: This comes from work done by Chris Berka and her team at Advanced Brain Monitoring. Berka gave a great TED talk about the work: https://www.youtube.com/watch?v=rBt7LMrIkxg.

113 Extreme athletes smooth out their heart rhythms: Author interview with Michael Gervais, 2016. A high-performance psychologist, Gervais was describing research done with Red Bull surfers.

114 "To diagnose . . . yourself while in the midst of action": Ron Heifetz, Marty Linsky, and Alexander Gradshow, *The Practice of Adaptive Leadership* (Boston: Harvard Business Press, 2009), p. 7.

Chapter Six: Pharmacology

116 They were hanging around with their noses:" Fiona Keating, "Dolphins Get High on a Diet of Toxic Fish," *International Business Times*, December 29, 2013.

116 Headlines like: Christie Wilcox, "Do Stoned Dolphins Give 'Puff Puff Pass' a Whole New Meaning?," *Discover*, December 30, 2013, http://blogs.discovermagazine.com/science-sushi/2013/12/30/stoned-dolphins-give-puff-puff-pass-whole-new-meaning/#.V_FInTKZNyo.

116 Psychopharmacologists have spent the past few decades: Ronald K. Siegel, *Intoxication: The Universal Drive for Mind-Altering Substances* (Rochester, VT: Park Street Press, 1989), p. 11

116 drug seeking and drug taking: Ibid., p. 99.

117 The carcasses of drunken birds: Ibid., p. 251.

117 The principle of conservation: Giorgio Samoroni, *Animals and Psychedelics: The Natural World and the Instinct to Alter Consciousness* (Rochester, VT: Park Street Press, 2002), p. 86.

118 "evolved to gratify our desires": Michael Pollan, "Cannabis, Forgetting and the Botany of Desire," *Occasional Papers of the Doreen B. Townsend Center for the Humanities*, no. 27 (2002).

118 "curiously, promote one plant": Ibid.

118 when Franciscan priests arrived in Mexico: Daniel Gade, *Spell of the Urubamba: Anthropogeographical Essays on an Andean Valley in Space and Time* (Cham, Switzerland: Springer, 2015), p. 208.

119 In Prohibition America: Michael Pollan, *The Botany of Desire* (New York: Random House, 2001), p. 9.

119 *Wired* dubbed him "Professor X": Ethan Brown, "Professor X," *Wired*, September 1, 2002.

119 the *New York Times* preferred "Dr. Ecstasy": Drake Bennett, "Dr. Ecstasy," *New York Times*, January 30, 2005.

119 "Gandalf" was not uncommon: Brian Vastag, "Chemist Alexander Shulgin, Popularizer of the Drug Ecstasy, Dies at 88," *Washington Post*, June 3, 2014.

119 Sasha Shulgin was born in Berkeley: For a great introduction to Shulgin's life, see the recent documentary *Dirty Pictures*, by Etienne Sauret, 2010.

120 "I learned there were worlds inside of me": James Oroc, "The Second Psychedelic Revolution Part Two: Alexander 'Sasha' Shulgin, The Psychedelic Godfather," Reality Sandwich, 2014, http://realitysandwich.com/217250/second-psychedelic-revolution-part-two/.

120 Sasha's interest," explains Johns Hopkins: Ibid.

121 The Shulgin Rating Scale: See https://en.wikipedia.org/wiki/Shulgin_Rating_Scale.

121 At 22 milligrams: Alexander Shulgin and Ann Shulgin, *PiHKAL: A Chemical Love Story* (Berkeley, CA: Transform Press, 1991), p. 560.

122 Richard Meyers, a spokesperson for the DEA: Bennett, "Dr. Ecstasy."

123 Everybody knows who the Shulgins: Teafaire, "No Retirement Plan for Wizards," teafaire.org, February 28, 2013, http://teafaerie.org/2013/02/456/.

124 Carhart-Harris didn't start out: Author interview with Robin Carhart-Harris, 2016.

125 the neurological impact of psilocybin: R. L. Carhart-Harris et al., "Implications for a Psychedelic-Assisted Therapy: A Functional Magnetic Resonance Imaging Study with Psilocybin," *British Journal of Psychiatry* 200 (2012): 238–44.

125 And the very first to explore LSD: Robin Carhart-Harris et al., "Neurol Correlates of the LSD Experience Revealed by Multimodal Imaging," *PNAS* 113, no. 17 (2016): 4853–58.

125 This is to neuroscience what the Higgs: Ian Sample, "LSD's Impact on the Brain Revealed in Groundbreaking Images," *Guardian*, April 11, 2016.

126 named Joseph Smith: For a great biography of Joseph Smith, see Richard Bushman, *Joseph Smith: Rough Stone Rolling* (New York: Vintage, 2007).

128 "[S]ome people can reach transcendent": Oliver Sacks, "Altered States: Self-Experiments in Chemistry," *New Yorker*, August 27, 2012.

128 Referring to Erowid: Erik Davis, "Don't Get High Without It," *LA Weekly*, April 29, 2004.

129 Take, for example, psychiatrist Rick Strassman's: Rick Strassman, *DMT: The Spirit Molecule* (Rochester, VT: Park Street Press, 2001).

129 "I had over twenty years in experience": Alex Tsakiris, "Dr. Rick Strassman on Whether Psychedelic Drugs Prove We Are More than Our Brain," Skeptico.com, March 25, 2016. This is a podcast: http://skeptiko.com/rick-strassman-psychedelic-drugs-prove-we-are-more-than-our-brain/.

130 The Lexicon is packed with neologisms: See https://wiki.dmt-nexus.me/Hyperspace_lexicon.

132 In 2010, chemist Lee Cronin: Author interview with Lee Cronin, 2016.

133 In August 2014, researchers at Stanford: Lexi Pandell, "Don't Try This at Home: Scientists Brew Opiates with Yeast," *Wired*, August 13, 2015.

133 The Canadian company HyaSynth Bio: Tracey Lindeman, "HyaSynth Bio Working to Mimic Medical Effects of Pot in Yeast," *Montreal Gazette*, June 29, 2015. Also see http://hyasynthbio.com.

133 This is really just the beginning: Author interview with Andrew Hessel, 2016.

Chapter Seven: Technology

135 For Dean Potter: For a deeper look at Dean's last flight, see Daniel Duane, "The Last Flight of Dean Potter," mensjournal.com, http://www.mensjournal.com/adventure/outdoor/the-last-flight-of-dean-potter-20150522. Also, Grayson Schaffer, "Dean Potter Killed in BASE-Jumping Accident," *Outside*, May 17, 2015.

136 "the *Rise of Superman* video series": If you'd like to view the entire six-part series, see http://www.flowgenomeproject.com/learn/videos/video-archive/.

136 "I know the dark secret": All the interviews with Dean were conducted between 2012 and 2015. There were many.

137 Steph Davis: Steph wrote a very powerful blog about her first husband's passing and her getting back in the air, See http://stephdavis.co/blog/dealing-with-death/ifly.

137 Thanks to innovators like Alan Metni: Author interview with Alan Metni, 2016.

138 Today, iFly: See author interview with Alan Metni, 2016, https://www.iflyworld.com.

139 Tony Andrews has been messing with your mind: For a really good overview of Andrew's life and his approach to sound and consciousness, he gave a comprehensive interview with *Mondo dr* that is now available in PDF form from Funktionone.com, http://www.funktion-one.com/dl/files/Tony%20Interview.pdf.

139 "I was feeling all this electricity": Tony Andrews, "Audio and Consciousness," a live interview conducted at PLASA London 2015, https://www.youtube.com/watch?v=D3RPJ8njrCY.

140 a common place to expand their minds: Ibid.

140 While studying the Arcy-sur-Cure caves: Steven Johnson, *How We Got to Now: Six Innovations That Changed the Modern World* (New York: Riverhead, 2015) and "Sound," *PBS: How We Got to Now*, November 12, 2014.

141 "Reznikoff's theory": Ibid., p. 88.

141 In Greece, churches: For an introduction to this work, see Josh Jones, "Mapping the Sounds of Greek Byzantine Churches: How Researchers Are Creating 'Museums of Lost Sound,'" Openculture.com, March 9, 2016.

141 In France, the Gothic arches of Notre Dame and Chartes: Kurt Blaukopf, *Musical Life in a Changing Society: Aspects of Music Sociology* (Portland, OR: Amadeus Press, 1982), pp. 180–82.

141 [I]n all societies," explains neurologist Oliver Sacks: Oliver Sacks, "The Power of Music," *Brain*, September 25, 2006, pp. 2528–32, http://brain.oxfordjournals.org/content/129/10/2528.

141 The burgeoning field of neuro-musicology: There's a ton of good information here, but three solid places to start are Rob-

errt Jourdain, *Music, the Brain, and Ecstasy: How Music Captures Our Imagination* (New York: William Morrow, 2008); Daniel Levitin, *This Is Your Brain on Music* (New York: Plume/Penguin, 2007); Jonah Lehrer, "The Neuroscience of Music," *Wired*, January 19, 2011.

141 Apple and the speaker manufacturer Sonos: *Fast Company* published an excellent overview of this research: John Paul Titlow, "How Music Changes Your Behavior at Home," *Fast Company*, February 10, 2016. Also see Mikey Campbell, "Apple Music and Sonos Launch a Collaborative Ad Campaign Touting the Benefits of Music," Appleinsider.com, February 10, 2016.

142 About forty-five years ago: Tony Andrews, Boom Festival, May 8, 2014, https://www.youtube.com/watch?v=q8xh6iZzMbk.

142 "What we're building in the Dance Temple": This was told to us by Android Jones in 2016.

143 Jones has established himself: The majority of the content in this section stems from a series of author interviews in 2016. For a good bio, Jones's website is the place to start: http://androidjones.com/about/bio/. If you want to go deeper, Jones has given a number of interesting interviews, such as "An Interview with Android Jones, the Digital Alchemist," Fractal Enlightenment, https://fractalenlightenment.com/35635/artwork/an-interview-with-android-jones-the-digital-alchemist.

143 On the Sydney Opera House: You can see a video of the entire performance at https://www.youtube.com/watch?v=e_ClOq0Wtkg.

146 In 2011, Mikey Siegel: Author interview with Mikey Siegel, 2016. Also, for an introduction to his work, see Mikey Siegel, "Enlightenment Engineering," TEDx Santa Cruz, May 1, 2014.

148 Consciousness Hacking: See http://www.cohack.life. Also see Noah Nelson, "Silicon Valley's Next Big Hack? Consciousness Itself," *Huffington Post*, March 3, 2015.

148 'Transformative Technology Conference: Siegel cofounded this conference with Dr. Jeffery Martin and Nicole Bradford. See: http://www.ttconf.org. Also Angela Swartz, "Meet the Transformative Technology Companies That Want to Help You Relax," Bizjournals.com, October 5, 2015.

148 A feature in *The New Yorker*: Nellie Bowles, "An evening with the Consciousness Hackers," *New Yorker*, June 23, 2015.

149 The Flow Dojo: For an overview: http://www.flowgenomeproject.com/train/flow-dojo/.

149 A prototype of the Dojo to Google's: If you want to see what all this looks like, see https://vimeo.com/153320792.

Chapter Eight: Catch a Fire

158 Burning Man aggressively extends the tradition of hedonic ecstasy: Lee Gilmore and Mark Van Proyen, ed., *AfterBurn: Reflections on Burning Man* (Albuquerque: University of New Mexico Press, 2005).

158 Michael Michaels': San Mateo City Innovation Week Panel discussion on the role of Burning Man in Silicon Valley culture, 2014, excerpt at https://vimeo.com/164357369 and entirety https://www.youtube.com/watch?v=b0yKsy-mWec.

159 You sink to your level of training: This quote is frequently attributed to an anonymous Navy SEAL (and often repeated by team members), but it most likely originated with the Greek poet Archilochus. "If you haven't been [to Burning Man]'": Nellie Bowles, "At HBO's 'Silicon Valley' Premiere, Elon Musk Has Some Notes," *ReCode*, April 3, 2014.

159 "So embedded, so accepted has Burning Man become": Vanessa Hua, "Burning Man," *SFGate*, August 20, 2000.

159 In 2013, John Perry Barlow, a fellow at Harvard Law School: @JPBarlow, Twitter.

160 Three years later, the actual president: "Just recently, a young person came up to me and said she was sick of politicians standing in the way of her dreams—as if we were actually going to let Malia go to Burning Man this year. Was not going to happen. Bernie might have let her go. Not us." White House Correspondents Association' Dinner, April 30, 2016, https://www.c-span.org/video/?c4591479/obama-drops-burning-man-joke.

160 In 2015, a team of scientists led by Oxford's Molly Crockett: Author interview May 12, 2016, and *Burning Man Journal*, http://journal.burningman.org/2016/05/black-rock-city/survive-and-thrive/researchers-share-first-findings-on-burners-transformative-experiences.

161 all combine to create a temporary autonomous zone: Hakim Bey, "The Temporary Autonomous Zone, Ontological Anarchy, Poetic Terrorism," http://hermetic.com/bey/taz_cont.html, anti-copyright, 1985, 1991.

161 "I like going to Burning Man": Will Oremus, "Google CEO

Is Tired of Rivals, Laws, Wants to Start His Own Country," *Slate,* May 15, 2013.

161 In 2007, Elon Musk did just that: Gregory Ferenstein, "Burning Man Founder Is Cool with Capitalism, and Silicon Valley Billionaires," *TechCrunch*, September 3, 2013.

161 He also came up with the ideas: Sarah Buhr, "Elon Musk Is Right, Burning Man Is Silicon Valley," *TechCrunch*, September 4, 2004; Ferenstein, "Burning Man Founder Is Cool with Capitalism, and Silicon Valley Billionaires."

161 Zappos founder and CEO Tony Hsieh: David Hochman, "Playboy Interview: Tony Hsieh," *Playboy*, April 2014.

162 While much has been made of the fact': Zack Guzman, "Zappos CEO Tony Hsieh Shares What He Would Have Changed About his $350M Downtown Las Vegas Project," CNBC, August 9, 2016, and Jennifer Reingold, "How a Radical Shift Left Zappos Reeling," *Fortune,* March 4, 2016.

162 As Burning Man cofounder Will Roger: Nellie Bowles, "Is Burning Man on the Cusp of Becoming a Permanent Utopian Community?," *New York* August 30, 2015.

163 This is all part of the evolution of Burning Man: "We Bought Fly Ranch," *Burning Man Journal,* June 10, 2016.

164 By the time the storm was over, it would spread $108 billion of damage: Mitigation Assessment Team Report, Federal Emergency Management Agency, https://www.fema.gov/media-library-data/20130726-1909-25045-8823/isaac_mat_ch1.pdf, p. 4.

164 "One of our guys took over a recon satellite": Bruce Damer interview, Joe Rogan Experience podcast, episode 561, October 14, 2014.

164 With Doctors without Borders as their inspiration: Daniel Terdiman, "Burn on the Bayou Showcases Burning Man Participants' post-Katrina Relief Efforts," CNET, March 24, 2008. Additional resources are the film *Burn on the Bayou*, Black Rock City LLC, 2008, and the organization's website, www.burnerswithoutborders.org.

165 Warner has a resume: Brian Calvert, "The Merry Pranksters Who Hacked the Afghan War," *Pacific Standard,* July 1, 2013. Dave Warner is a fascinating character, brilliant, iconoclastic, and highly productive. See also Andreas Tzortzis, "Learning Man," *Red Bull Bulletin,* August 2014.

166 "I'm dismantling the Death Star": Ibid.

166 With so much experience in self-organizing": Peter Hirshberg, *From Bitcoin to Burning Man and Beyond* ([N.p.]: Off the Common Books, 2014).

167 It's for this reason that Rosie von Lila: Author interview with Rosie von Lila, July 25, 2016.

167 Burning Man demonstration projects": Washoe Tribe installing solar panels at seven sites," *Record-Courier*, June 8, 2015; "Stained Glass 'Space Whale' to Blow Minds at Burning Man," *Reno-Gazette Journal,* November 13, 2015; "How a Chat App for Burning Man Turned into a Tool for Revolution," *AdWeek*, March 25, 2015.

168 Burning Man didn't invent the festival: Hirshberg, *From Bitcoin to Burning Man and Beyond.*

168 In Europe, we saw this: Jonathon Green, *Cannabis* (New York: Pavilion Press, 2002).

168 In the 1920's socialite Mabel Dodge Luhan's Taos home: Lois Palken Rudnick, *Utopian Vistas: The Mabel Dodge Luhan House and the American Counterculture* (Albuquerque: University of New Mexico Press, 1998).

168 In the 1960's Esalen's founders and faculty: Jeffrey Kripal, *Esalen: America and the Religion of No Religion* (Chicago: University of Chicago Press, 2008).

169 The power of shared peak experiences: See http://future.summit.co

169 "We wanted a permanent home": Author interview with Jeff Rosenthal, June 21, 2016.

170 The series, which has been called "TED crossed with Burning Man": Andy Isaacson, "Summit Series: TED Meets Burning Man," *Wired*, February 27, 2012; Steven Bertoni, "Summit Series Basecamp: The Hipper Davos," *Forbes,* January 26, 2012.

171 MaiTai Global: Author interview, August 22, 2016.

172 "We curate our experiences very strategically": Author interview with Jeff Rosenthal, August 15, 2016.

172 $20 billion": Richard Godwin, "How to Network like the One Percent," *Sunday Times,* June 18, 2016; Kim McNichols, "Kiteboarding Techies Generate $7 Billion in Market Value," *Forbes*, December 7, 2011.

173 We experienced this firsthand: Jamie Wheal, "Five Surprising Ways Richard Branson Harnessed Flow to Build a Multi-Billion Dollar Empire," *Forbes,* March 25, 2014.

173 Carbon Warroom,: See: www.carbonwaroom.com.

173 "That's where I had the idea for Virgin Galactic": Wheal, "Five Surprising Ways."

174 In his seminal book: Geoffrey Moore, *Crossing the Chasm: Marketing and Selling High-Tech Products to Mainstream Customers* (New York: Harper Business, 2006); Everett Rogers, *Diffusion of Innovation* (New York: Free Press, 2003).

175 Eighteen million Americans now have a regular practice: T. C. Clarke et al., "Trends in the Use of Complementary Health Approaches Among Adults: United States, 2002–2012," National Health Statistics, No. 79, Hyattsville, MD, National Center for Health Statistics, 2015; "Uses of Complementary Health Approaches in the U.S.," National Center for Complementary and Integrative Health.

175 44 percent of all U.S. companies will offer mindfulness: "Corporate Mindfulness Programs Grow in Popularity," National Business Group on Health and Fidelity, July 14, 2016.

175 Since rolling out their program, Aetna estimates: Joe Pinsker, "Corporations' Newest Productivity Hack: Meditation," *Atlantic*, March 10, 2015.

175 the meditation and mindfulness industry grew to nearly $1 billion: Jan Wieczner, "Meditation Has Become a Billion-Dollar Business," *Fortune*, March 12, 2016.

175 At Harvard, Professor Tal Ben Shahar's: Craig Lambert, 'The Science of Happiness," *Harvard Magazine*, January–February 2007.

175 By college, many Millennials have reached: Pfaffenberger, ed., *The Postconventional Personality*, p. 60.

175 researchers began finding the practice did everything: N. P. Gothe and E. McAuley, "Yoga and Cognition: A Meta-Analysis of Chronic and Acute Effects," *Psychosomatic Medicine* 77, no. 7 (September 2015): 784–97; N. R. Okonta, "Does Yoga Therapy Reduce Blood Pressure in Patients with Hypertension? An integrative Review," *Holistic Nursing Practitioner* 26, no. 3 (May–June 2012): 137–41.

175 As of 2015, some 36 million Americans: Marlynn Wei, "New Survey Reveals the Rapid Rise of Yoga—and Why Some People Still Haven't Tried It," *Harvard Health Publication*, June 15, 2016.

176 more popular, in terms of participation, than football: 2016

Yoga in America Study, *Yoga Journal* and Yoga Alliance, http://www.yogajournal.com/yogainamericastudy/.

176 Bulletproof has grown into a nine-figure company: Author interview with Dave Asprey, 2015.

176 the fastest-growing industry: Will Yakowicz, "Legal Marijuana Blooms into the Fastest Growing Industry in America," *Inc.*, January 27, 2015.

176 The whole of the cannabis economy: *The State of Legal Marijuana Markets,* 4th "ed., Arcview Market Research, 2016.

176 As Peter Reuter: Eliott McLaughlin, "As Haze Clears, Are Americans' Opinions on Marijuana Reaching a Tipping Point?," CNN, August 30, 2013.

176 Thirty-two million Americans use psychedelics: T. S. Krebs P. Ø. Johansen, "Psychedelics and Mental Health: A Population Study," *PloS One*, August 13, 2013.

176 a 2013 study: Ibid.

177 fewer than four hundred patents for neurotech: "New Gold Rush for US Patents: Brain Technologies," *TRTWorld*, May 7, 2015, http://www.trtworld.com/business/new-gold-rush-us-patents-brain-technologies-819.

178 "They stay up all night dancing": Harold Bloom, *The American Religion* (New York: Touchstone, 1992), p. 59. Bloom's book is a fascinating contexualization of American religiosity, with an emphasis on tracing the unique through-lines of the direct and experiential traditions. For additional scholarly assessments, see Jon Butler, *Awash in a Sea of Faith* (Cambridge, MA: Harvard University Press, 1990), and Nathan Hatch, *Democratization of American Christianity* (New Haven, CT: Yale University Press, 1989).

Chapter Nine: Burning Down the House

182 It was 1953 and the Pentagon had a problem: Annie Jacobsen, *The Pentagon's Brain* (New York: Little, Brown, 2015), p. 103.

182 So the Secretary of Defense demanded: Ibid, p. 104.

182 But the CIA had been discreetly testing: Ibid, p. 105.

183 got wind that a brilliant young neuroscientist: John Lilly, *The Scientist* (Berkeley, CA: Ronin, 1988), p. 90.

183 Lilly had solved the two biggest technical problems of mechanically inducing ecstasis: Ibid, pp. 87–88.

183 In primates, Lilly had discovered: Ibid, p. 90.
183 "Anybody with the proper apparatus": Ibid, p. 91.
184 To guard against this, Lilly detailed: Ibid, p. 92.
184 Not long after that initial presentation: Ibid, p. 93.
184 A few years later, *Harper's Magazine*: Ibid, p. 96.
185 History shows a typical progression of information technologies: Tim Wu, *The Master Switch* (New York: Knopf, 2010), p. 6.
185 All these disconnected communities and houses will be united through radio: Ibid., p. 37; *Bradford Science and Technology Report*, No. 8, August 2007.
186 Before any question of free speech: Ibid., p.13.
186 As W. B. Yeats put it: This is apocryphally attributed to Yeats, who was an initiate in several mystery cults; it's also as likely a variant on Eden Phillipot's 1919 book, *A Shadow Passes:* "The universe is full of magical things patiently waiting for our wits to grow sharper."
186 consider that elite athletes": Matt Slater, "Has the Biological Passport Delivered Clean or Confused Sport?," *BBC Sport*, November 12, 2014. This is really only the tip of the iceberg. As ethnopharmacologist Jon Ott has noted, the Controlled Substances Act of 1970 specifically prohibits any possession of DMT in any amount, which, given that it is produced endogenously in humans, means that "any human being is guilty of such possession." In Graham St. 'John, *Mystery School in Hyperspace: A Cultural History of DMT* (Berkeley, CA: Evolver, 2015), p. 8.
187 slip Fidel Castro an LSD-soaked cigar: Fabian Escalante, *Executive Action: 634 Ways to Kill Fidel Castro* (Ocean Press, 2006).
187 "Within the CIA itself, [agents] were taking LSD regularly" Jay Stevens, *Storming Heaven: LSD and the American Dream* (New York: Grove Press, 1998), p. 82.
188 a chemist at Fort Detrick's Biological Weapons Center: James Rissenov, "Suit Planned Over Death of Man C.I.A. Drugged," *New York Times*, November 26, 2012.
188 Stegner dismissed him as a sort of highly talented illiterate: Jackson J. Benson, *Wallace Stegner: His Life and Work* (Lincoln: University of Nebraska Press, 1996), p. 253.
188 "The scientists didn't have the guts": Joshua Fried, "What a Trip," *Stanford Alumni Magazine*, January/February 2002.
189 "Volunteer Kesey gave himself over to science": Tom Wolfe,

The Electric Kool-Aid Acid Test (New York: Farrar, Straus & Giroux, 1968), p. 45.

189 Half the time": Ibid., p. 46; Richard Strozzi-Heckler, *In Search of the Warrior Spirit: Teaching Awareness Disciplines to the Green Berets* (Berkeley: Blue Snake Books, 2007), p. 17.

189 Armed with speakers mounted in the redwoods: John Markoff, *What the Doormouse Said* (New York: Viking, 2005), p. 122.

190 A round of post-Vietnam soul-searching": FrankRose, "A New Age for Business?," *Fortune,* October 8, 1990.

190 "I just made it my weekend duty": Jim Channon, interview, *Goats Declassified: The Real Men of the First Earth Battalion* (Anchor Bay Entertainment, 2009).

190 He penned: Jim Channon, *First Earth Battalion Operations Manual* ([N.p.]: CreateSpace, 2009), p. 64.

190 "Beam Me Up Spock": John Alexander, "Beam Me Up Spock: The New Mental Battlefield," *Military Review,* December 1980.

191 Almost as an afterthought': Channon, *First Earth Battalion Operations Manual* (CreateSpace, 2009), p. 66.

191 In May 2003, *Newsweek*: Adam Piore, "PSYOPS: Cruel and Unusual," *Newsweek*, May 19, 2003; Alex Ross, "When Music Is Violence," *New Yorker*, July 4, 2016.

192 In heavily redacted documents: Seth Richardson, "Vegas FBI Investigated Burning Man in 2010," *Reno-Gazette Journal,* September 4, 2015. Muckrock is the organization that originally published the findings: https://www.muckrock.com/news/archives/2015/sep/01/burning-man-fbi-file/.

192 More likely, the FBI was taking a page: David Cunningham, *There's Something Happening Here: The New Left, the Klan, and FBI Counterintelligence* (Berkeley: University of California Press, 2004).

192 And while it's hard to tell if it's an anomaly: Sarah Maslin Ner, "Burning Man Ends, and an Event for Law Enforcement Begins," *New York Times,* September 11, 2015. Original citation in *Reno Gazette-Journal* is no longer retrievable, cited at https://burners.me/2013/08/23/pershing-county-cops-and-federal-agents-integrated-and-synchronized/.

193 In 2007, a collection of the world's biggest brands: Martin Lindstrom, *Buyology: The Truth and Lies About Why We Buy* (New York: Crown Business, 2010), p. 12.

194 discovered that product placement: Ibid., p .14.

194 no discernible way': Ibid., p. 126.
195 At the tail end of the twentieth century: Alvin Toffler, *Future Shock* (New York: Bantam, 1984), p. 221.
195 It's how Starbuck's: Matthew Dollinger, "Starbucks, 'The Third Place,' and Creating the Ultimate Customer Experience," *Fast Company*, June 11, 2008.
195 were at the Advertising Research Foundation: B. Joseph Pine II and James H. Gilmore, *The Experience Economy* (Boston: Harvard Business Review Press, 2011), p. 255.
195 You may or may not get as lean as those models: Thu-Huong Ha, "New Ads from Equinox Show Gym-goers at Peak Absurdity," *Quartz*, January 13, 2016.
196 That's a positive transformation": For a fascinating take on "reverse brands" like CrossFit and Ikea, which strip out presumed amenities in favor of delivering on a different set of consumer desires, see Youngme Moon', *Different: Escaping the Competitive Herd* (New York: Crown Business, 2010).
196 Consider a recent Jeep campaign: Presentation at Advertising Research Foundation by Omnicom agency responsible for campaign. Additional details at http://m.jeep.com/jeep_life/news/jeep/stick_in_the_mud.html and http://media.fcanorthamerica.com/newsrelease.do?id=1919&mid=46.
196 a multibillion-dollar industry that employ the best neuroscientists: Seth Ferranti, "How Screen Addiction Is Damaging Kids' Brains," *Vice*, August 6, 2016.
197 In their study, a trained storyteller: Author interview with Chris Berka, Advanced Brain Monitoring, the company responsible for conducting the study, January 27, 2015.
198 It's very easy to imagine: Kevin Kelly, "The Untold Story of Magic Leap, the World's Most Secretive Startup," *Wired*, May 2016.
198 This comprehensive tracking": Ibid.
199 "all the advantages of Christianity and alcohol; none of their defects": Aldous Huxley, *Brave New World* (London: Chatto & Windus, 1932), p. 54.
199 "In *Brave New World,* they are controlled by inflicting": Neil Postman, *Amusing Ourselves to Death* (New York: Penguin, 2005), p. viii.
199 all power that derives from the control of information: Tim Wu, *The Master Switch* (New York: Knopf, 2010), p. 310.
200 The Cycle is powered by disruptive innovations: Ibid., p. 20.

Chapter Ten: Hedonic Engineering

201 Sasha Shulgin used to say:" *Dirty Pictures.*

202 First identified back in the 1930s, Jerusalem Syndrome: Yair Bar-El et al., "Jerusalem Syndrome," *British Journal of Psychiatry* 176 (2000): 86–90.

203 It's why Burning Man advises people: This essay is one the more frequently reposted and entertaining articles on "decompression" or coming back to regular life after the event, by "The Colonel" of Arctic Monkey Camp: "Do Not Divorce Your Parakeet Yet," *New York Burners Guide*, original date and publishing location unknown. Here's a look at the first paragraph:

"DO NOT MAKE CHANGES TO YOUR LIFE FOR AT LEAST THREE WEEKS AFTER YOU COME BACK FROM BURNING MAN.

"Do not quit your job. Do not divorce your wife, husband, sister, dog, parakeet. Do not sell all your possessions and move to Tibet to be a monk. Do not ditch your car and travel the world. Do not found Hobbit Camp. Do not plan a giant zeppelin for next year's Burn. Do not move out of your house, break up with your girlfriend, boyfriend, get married, move in your playa lover, sell your car, ditch your friends, or make other rash decisions after you come home. This is important, because the playa is still going to be in your brain, and the effects are like that of rarefied stupid sometimes. It will make total sense to have a threesome with your significant other and someone in an enormous rabbit costume at the Burn; in reality the ears get caught in the ceiling fan. Make sure if you have major life decisions to make, you make them AFTER you settle down and settle in. The emotions and the stress will still be in your system for some time; do not allow them to unduly influence your life."

203 In 2009, Swiss neurologist Peter Brugger: Peter Brugger, Christine Mohr, Peter Krummenacher, and Helene Haker, "Dopamine, Paranormal Belief, and the Detection of Meaningful Stimuli," *Journal of Cognitive Neuroscience* 22, no. 8, 2010: 1670–81.

203 When the prefrontal cortex shuts down, impulse control: Julie A. Alvarez and Eugene Emory, "Executive Function and the Frontal Lobes: A Meta-Analytic Review," *Neuropsychology Review* 16, no. 1 (March 2006). We have oversimplified the relationship between the PFC and executive function; as this

meta-analysis suggests, there are more complex relationships between neuroanatomy and consciousness.

203 As Buddhist teacher and author Jack Kornfield: Jack Kornfield, *After the Ecstasy, the Laundry: How the Heart Grows Wise on the Spiritual Path* (New York: Bantam, 2001).

203 In 1806, General Zebulon Pike: Zebulon Pike, *Account of Expeditions (1806–1807)"* (Madison: Wisconsin Historical Society, 2003).

204 Our ability to accurately estimate how close things are to happening: Burkhard Bilger, "The Possibilian"," *New Yorker*, April 25, 2011. David Eagleman is a friend and advisory board member, and doing some of the more interesting work on time perception. This *New Yorker* article is a great introduction to some of the basics in his research.

204 Popular religious movements, from the Seventh-Day Adventists: Leon Festinger, *When Prophecy Fails: A Social and Psychological Study of a Modern Group That Predicted the Destruction of the World* (New York: Harper Torchbooks, 1964).

204 Contemporary psychonauts have even coined a term: https://wiki.dmt-nexus.me/Hyperspace_lexicon#Eschatothesia.

205 "Most people overestimate": Bill Gates, *The Road Ahead* (New York: Viking Press, 1995), p. 316 (this is the closest documented source we can find for the more folksy variant we quote).

205 In 2014, Ryan Holiday: Ryan Holliday, The Obstacle Is the Way: The Timeless Art of Turning Trials into Triumph (New York: Portfolio, 2014), p. xiv.

206 Nick Mevoli: Adam Skolnick, "A Deep-Water Diver From Brooklyn Dies After Trying for a Record," *New York Times*, Nov. 17, 2003.

206 Water is acceptance: Nick Mevoli, "How I Got to 91 Meters," freediveblog.com, See: http://www.freediveblog.com/2012/06/11/how-i-got-to-91-meters-by-nick-mevoli/.

206 The biggest problem with freedivers: "Blue Hole, Black Hole," *Economist*, Feb. 27, 2016.

207 One with the World, Skolnick, ibid.

207 On one occasion the tank was too warm: Lilly, *The Scientist*, p. 158. All subsequent information from this section drawn from this and related autobiographical reports.

209 "No sympathy for the devil": Hunter S. Thompson, *Fear and Loathing in Las Vegas* (New York: Vintage, 1998), p. 89.

209 If the North Face': Hans Ludwig, "The Return of the Extreme Skier," *Powder*, December 2010.

209 "We got pinned on a 70-degree face": Kristen Ulmer, author interviews, June 15–21, 2016. All subsequent information in this section is drawn from same.

212 The meeting was dedicated to resisting Calvinistic attacks: Michael Moss, *Salt Sugar Fat: How the Food Giants Hooked Us* (New York: Random House, 2014), p. 10. All subsequent quotes on this topic are from here.

213 as UCLA's Ron Siegel suggests: Siegel, *Intoxication*, p. 209.

215 Research shows we're more likely to keep habits: Katherine Milkman, "The Fresh Start Effect: Temporal Landmarks Motivate Aspirational Behavior," Wharton School Research Paper No. 51, December 24, 2013.

216 "The road of excess": William Blake, *The Marriage of Heaven and Hell* (Benedictine Classics, 2010), p. 11. This general observation leads to the notion of left-hand and right-hand paths to knowledge. The right hand are the orthodox paths geared for the lowest common denominators, full of "Thou Shalts" and "Thou Shalt Nots." Imagine techniques of ecstasy designed by lawyers and bureaucrats. The left-hand path (of which tantra is one of the better-known examples, but also including Western SexMagik and other pursuits) seeks to embrace all the most distracting and addicting pursuits—sex, drugs, rock and roll—among them, to get to realization faster. The left-hand path is, arguably, the fastest path to radical awakening, but also the one with the lowest completion rate. That is why we are advocating a "middle path" here that includes permission to explore ecstatic states with the flexible "liberating structures" built into hedonic calendaring. As far as we know, we haven't seen this kind of synthesis offered before and hope it's useful for current explorers.

216 As Hemingway reminds us: Ernest Hemingway, *A Farewell to Arms* (New York: Scribner, 2014), p. 318.

216 "[It's] a widespread tendency": John Welwood, *Toward a Psychology of Awakening* (Boston: Shambhala, 2002), p.5.

217 "Embracing our vulnerabilities is risky but not nearly as dangerous as giving": Brené Brown, "Power of Vulnerability," TEDx Houston, June 2010.

217 "Love tells me I am everything": Nisagardatta, *I Am That*

(Durham, NC: Acorn Press, 2012). There are lots of variants on this quotation; we have chosen the one that seemed most descriptive.

217 "[Ecstasis] is absolutely ruthless and highly indifferent": John Lilly, *Dyadic Cyclone: The Autobiograpghy of a Couple* (New York: Pocket, 1977). This is, as best we can tell, the original source for a quote of Lilly's that has been widely requoted elsewhere. We took the small liberty of replacing Lilly's "Cosmic love" with the word *ecstasis* here for continuity of terms—believing that both refer to comparable experiences of selfless information richness in altered states.

218 The Japanese get at this same idea: Leonard Koren, *Wabi-Sabi for Artists, Designers, Poets and Philosophers* (Berkeley, CA: Stone Bridge Press, 1994), p. 67.

218 "Ring the bells that still can ring": Leonard Cohen, "Anthem," *The Future*, Sony Music, 1992.

Conclusion

219 So the founder of Oracle Corporation: Stu Woo, "Against the Wind: One of the Greatest Comebacks in Sports History," *Wall Street Journal*, February 28, 2014. All factual references to the race in this section drawn from here plus author interview/conversation with James Spithill at Red Bull's Glimpses Conference, June 2014.

221 Epimetheus, whose name means "afterthought": Robert Graves, *The Greek Myths* (New York: Penguin, 1993), p. 148.

222 "That is why": Aesop, Laura Gibbs, *Aesop's Fables* (Oxford: Oxford University Press, 2008), p. 242.

INDEX

ABOUT THE AUTHORS

STEVEN KOTLER is a *New York Times* bestselling author, award-winning journalist, and the cofounder and director of research for the Flow Genome Project. His books include *Tomorrowland, Bold, The Rise of Superman, Abundance, A Small Furry Prayer, West of Jesus,* and *The Angle Quickest for Flight.* His work has been translated into forty languages and his articles have appeared in more than eighty publications, including the *New York Times Magazine, Atlantic Monthly, Wired, Forbes,* and *Time.* Steven is an in-demand speaker and advisor on technology, innovation and peak performance. You can find him online at www.stevenkotler.com.

JAMIE WHEAL is a world-leading expert on peak performance and leadership, specializing in the neuroscience and application of flow states. He has advised everyone from the U.S. Naval War College and Special Operations Command, to the athletes of Red Bull, to the executives of Google, Deloitte, and Young Presidents' Organization, to the owners of the top professional sports teams in the United States and Europe. Jamie is a sought-after speaker and adviser to top performers across disciplines, and his work has appeared in anthologies and peer-reviewed academic journals.